CW00546826

AMBER

Tears of the Gods

AMBER

Tears of the Gods

NEIL D. L. CLARK

Hunterian Museum and Art Gallery, University of Glasgow

DUNEDIN

Published by
Dunedin Academic Press Ltd
Hudson House
8 Albany Street
Edinburgh EH1 3QB
Scotland

www.dunedinacademicpress.co.uk

ISBN 978-1-906716-16-5 (Hardback)
ISBN 978-1-906716-17-2 (Paperback)

British Library Cataloguing in Publication Data
A catalogue record for this book is available from the British Library

Design and production by Makar Publishing Production, Edinburgh
Printed and bound in Poland by Hussar Books

For Alex and Catriona

Pretty! In amber to observe the forms
Of hairs, or straws, or dirt, or grubs, or worms,
The things, we know, are neither rich nor rare
But wonder how the devil they got there?

From: Alexander Pope's *Epistle to Dr Arbuthnot* (1735)

Amber nugget from the Collections
of the Hunterian Museum,
University of Glasgow

Contents

Amber fishermen using nets in stormy winter seas off the Baltic Coast.

Introduction

In 2010, a unique exhibition of amber art and artefacts arrived in Glasgow, mostly coming from the famous Malbork Castle collection in Poland. Notable additions included the well-known Gdańsk lizard from the Gdańsk Amber Museum, as well as a selection of insects trapped in amber and some historical artefacts from the collections of the University of Glasgow's Hunterian Museum. The aim of the exhibition was to bring together many different aspects of amber that would not normally be exhibited together in such a holistic way. Amber collected for its historic and prehistoric significance and those pieces collected for natural history reasons were displayed together to help us to understand amber in its entirety. This book draws on the exhibition for inspiration, looking at many aspects of amber from prehistory to natural history; at how people related to amber from the Stone Age, through the Middle Ages until the present day.

Amber is used around the world, sometimes for medical or spiritual wellbeing, sometimes for adornment or decoration, and sometimes for scientific reasons. Wherever amber is found, it is treasured by its captors who, in turn, are captivated by its wonder. Amber comes in many varieties, colours, and forms. Amber from the Baltic region of Europe is one of the most abundant in the world and has been used by man for over 40,000 years. Although Czech amber is one of the earliest to be used in Europe, the more northern Baltic amber started to become the main source available to ancient peoples as rapidly as the ice sheets of the last ice age retreated 10,000 years ago.

Baltic amber has been given different names derived from its religious or spiritual value and its physical properties. The ancient Greeks called it ηλεκτρον (*electron*) after their Sun God, but the Romans, who

Amber goblet from the 18th century collections of Dr William Hunter in the Hunterian Museum, Glasgow.

Woodcut by Jacob Meydenbach of the amber tree (1491).

better understood the origins of amber, gave it the more practical name of *sucinum* which means 'juice' or 'sap' of a tree. In the northern part of Europe, there are two common names for amber; one is the Germanic name *Bernstein* (burning stone) and its variants, and the other is based on the name used in the eastern Baltic states, *Yantar* and its variants, which are supposed to be derived from the Latin verb *ientare* (to have breakfast). Perhaps this relates to the time of sunrise? One of the early words used for amber, recorded by the Romans in the first century AD, was that used by the Aisti tribe of the Baltic region who called amber *Glaesum*, which probably gave rise to the English word *glass*.

The origins of the word 'amber' are not clear. Some have suggested that it may be from the Medieval Latin word *ambrum* (scent) which was derived from the Arabic word *anbar* which is also related to scent and possibly brought back to Europe by the Crusaders of the eleventh to thirteenth centuries. The Arabic word *anbar* came to be used for amber and what is now known as ambergris. Ambergris is a gray resinous product from the intestine of a sperm whale, often found floating in the sea. Amber and ambergris have both been burnt as incense, but to differentiate between the two, the French used the terms *ambre gris* (gray amber) and *ambre jaune* (yellow amber) adopted respectively as 'ambergris' and 'amber' in English.

Many other suggestions have been made on the origins of the term, but it seems possible that it derived from the Phoenician word for amber *Yiaintar*. The Phoenicians traded with the Baltic amber-producing nations over 3,000 years ago, possibly taking it to the countries along the sea routes into the Mediterranean. Certainly the nations along the eastern trade route from the eastern Baltic to the Black Sea all have variations of the Russian янтарь (*Yantar*). The similarity between the terms *Yantar* and the Arabic word *anbar* suggests perhaps that the word has changed only slightly over the centuries.

The German word for 'to burn' is *anbrennan* and in Belgium the ancient word for amber is *Anbernen*, suggesting that it is also possible that the word 'amber' may come from the Germanic word describing what the stone was used for — for burning as an incense.

The extraction of amber from the Baltic dates back to prehistoric times. People in the region have collected amber by different means over the millennia: by collecting flotsam from the beaches, fishing with nets in the sea, dredging the rivers, and quarrying the blue clay in which the amber may be found on land. Steeped in culture and tradition, since early times amber was a highly valued commodity, often more so than gold.

As with most treasured gems, the history of those who had the amber, and those who wanted it, is dotted with battles and wars to control the trade of this valued solar jewel.

Name for amber	Language	Name for amber	Language	Name for amber	Language
Amber	English	Yantar	Russian	Bernstein	German
Alambre	Portuguese	Jantar	Polish	Barnsten	Swedish
Ambar	Spanish	Yainitar	Phoenician	Barnsteen	Dutch
Ambra	Italian	Dzintars	Latvian	Burshtinen	Yiddish
Ambre	French	Gintaras	Lithuanian	Borostyan	Hungarian
Anbernen	Ancient Belgian	Chihlinbar	Romanian	Bursztyn	Polish

Some of the many different names for amber based on the three common names used in the Baltic region and Western Europe.

Amber origins

BALTIC AMBER AND THE NATURE OF AMBER

Although there have been many interpretations over the centuries as to exactly what amber is, it was suggested as far back as 600 BC that it was resin of a tree. Amber is now universally known to be a fossilised resin that once flowed from the trunks and boughs of ancient, but now extinct, trees. The most famous of all ambers is that found in the region of the Baltic Sea. The trees that produced the resin lived about 40 million years ago (mya) during the Eocene Epoch. This organic jewel is sometimes incorrectly referred to as tree sap. Tree sap is the fluid that circulates through a plant's vascular system, while resin, a semi-solid amorphous organic

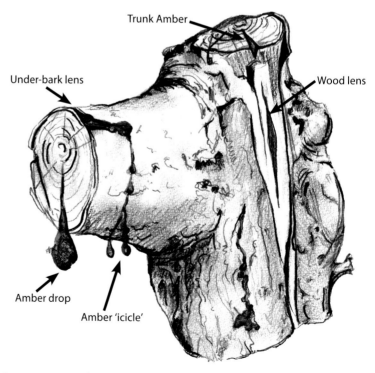

Representation of a tree producing different amber morphologies.

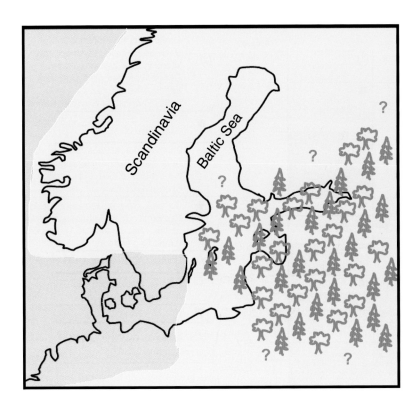

The likely extent of the amber forest (green trees) during the Eocene Period with land shown in yellow.

substance, flows from special ducts of some plants, and is especially seen flowing from conifers, but also found in deciduous trees. The resin is used by these plants to protect them from attacks by parasitic fungi, insects, mechanical damage, and even extreme cold weather.

Baltic amber comes from the ancient forests of the Baltic region, which extended from land that is now partly submerged beneath the northern European Baltic Sea, and may have originally extended as far south as the Black Sea between south-eastern Europe and Asia. The climate here at that time was warm, humid, and subtropical with tall tropical trees and a lush undergrowth of ferns and moss; a forest very different to the coniferous boreal forests and tundra landscape of the region today. The trees grew at a time when the planet was undergoing a global warming event that lasted from about 60 million years ago until about 30 million years ago when cooling took place, turning forest into open countryside. When the Baltic resins flowed from these extinct trees, the forest was much more like that of tropical parts of southern Asia. Evidence from fragments of plants held within the amber resin suggests that the forest was rich in species. Oak flowers, pine needles, cypress leaves and twigs, cedar leaves, moss and many more plant remains can all be found as inclusions in Baltic amber.

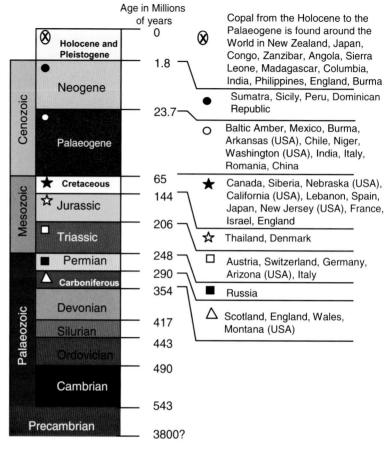

Geological timescale with major amber and copal occurrences.

Baltic amber, however, is not unique. Fossilised tree resin is known from rocks as old as over 310 million years and perhaps even older. Most resins that are commonly called amber come from the Mesozoic era (248–65 million years ago) or Cenozoic era (65–1.8 million years ago). Rare 'nuggets' of amber from the Cretaceous period, over 65 million years ago, contain fossilised insects that were alive and flying as dinosaurs walked the Earth. Although Jurassic period amber is known, insects have yet to be found in Jurassic amber. Very rare inclusions of plant and fungal remains in amber are known from pieces as old as the Carboniferous period, over 300 million years ago. The species of tree that produced amber is unknown, but researchers have attempted to identify it from the chemistry and properties of the amber, the results of which are still inconclusive. Different ambers from other parts of the world appear to have been produced by different types of resin-producing plants. The term 'amber' is taken by some to apply only to amber originating in the Baltic region, due to its unique composition, which includes an organic compound called succinic acid. However, the word 'amber' has become a term applied to most fossilised resins that have similar properties.

Amber is particularly abundant from the Cenozoic era of Central and North America, Europe, and Asia. Particularly famous is the amber from the Dominican Republic (less than 40 million years old) and Mexico (about 30 million years old). These ambers are notably different from Baltic amber, due to the greater abundance of plants and animals trapped within them, and to their transparency. However, Baltic amber, having a long history of trade going back thousands of years to the Stone Age, still dominates the world amber market.

Political map of the Baltic region.

Baltic amber is most commonly found in the coastal regions of Denmark, northern Germany, southern Sweden, Poland, Russia, Lithuania, Latvia, Estonia, and Finland, but can also be found in glacial and sedimentary deposits quite a distance inland. These rich deposits of amber have been exploited for many reasons over the millennia. Amber was highly prized by Sun-worshipping peoples, such as the Phoenicians, the Romans and the Greeks, who governed the populations of the European world in ancient times. Amber has been used as a talisman against sickness and bad luck, and as a cure for many illnesses. It was used as an adornment by the wealthy and carved into religious symbols for temples and churches of various religions. Today it is still used as a valued gemstone in jewellery and as a healing stone by some. To the scientist, it is a source of information on a multitude of topics from climate change to plant and animal taxonomy; from medicine to static electricity.

RESINOUS PLANTS

Some living trees produce copious amounts of resin, much as we might imagine the extinct amber trees of the Baltic to have done. One present-day tree, the Kauri tree (or araucarian trees of the genus *Agathis*) found

Left: Modern millipede trapped in larch resin. *Right:* Resin flowing on the bark of a modern larch tree.

in the nearby islands of New Zealand and in south-east Asia, produces vast amounts of resin and has done so for centuries.

The bark of the *Boswellia* tree, growing from North Africa to India, is scraped to produce a resin in the form of little lumps called 'tears'. The resin is commonly used in the perfume industry and is more commonly known as 'frankincense'. Frankincense is also used in medicine, in the form of an essential oil, or a powder, to treat a diverse range of illnesses including Crohn's Disease and arthritis.

Myrrh is also a tree resin that is derived from a number of species, but most commonly from the *Commiphora myrrha* tree from eastern Africa and the Arabian Peninsula. It has been used since the time of the pharaohs in Egypt until about the fifteenth century to embalm the dead, and also as a medicine to cure illness. It has been used historically and is still used today in a similar way to frankincense to treat arthritic conditions, as well as in the treatment of gum diseases. Frankincense has been a valuable commodity traded on the Arabian Peninsula and in North Africa for more than 5,000 years. Frankincense and myrrh, along with copal and other tree resins, are used as incense due to their pleasant and highly aromatic scent. Pellets of incense containing frankincense were found in the tomb of the ancient Egyptian King Tutankhamun, who died over 3,000 years ago. Charred frankincense was ground into a powder called 'kohl' by the Egyptians, which gave us the name for coal, and is still used by some traditional communities today. It was used as a cosmetic product to make the distinctive black eyeliner, as seen on so many figures in Egyptian art and wall paintings.

THE BALTIC AMBER TREE

The identity of the amber-producing tree of the ancient forests is unclear. It may even be that there was more than one type of tree producing amber. Early attempts based on inclusions and the presence of succinic acid in amber indicated an affinity with pine trees. The amber-producing tree from the Baltic had been named *Pinus succinifera*, despite no identifiable amber-producing part of the tree having been found. However, pieces that have been found suggest that the tree was not of the genus *Pinus* at all, but something quite different. As a result, it has been suggested that the tree should be given the new name *Pinites*, rather than *Pinus*, since it was only known from fossil remains and was quite different from the living *Pinus*. A detailed comparison of the chemistry of Baltic amber with other living resin-producing trees suggests that the amber may have come from an araucarian relative (the same group as the Monkey Puzzle) of the pine tree very similar to *Agathis* (the Kauri gum tree). However, no araucarian remains have been found as inclusions in Baltic amber. Some researchers suggest that the amber-producing tree may have been an extinct ancestor to both the present-day pines and araucarians. It may have had the physical appearance of a pine, and a resin with a similar chemical composition to an araucarian. Others have suggested that it might have been an extinct cedar or larch that produced Baltic amber. In other words, we are uncertain which type of tree produced the vast quantities of Baltic amber we find today.

The modern araucarian Monkey Puzzle tree.

THE COLOUR OF AMBER

One method of valuing Baltic amber is by its colour. Amber can be transparent right through to completely opaque, and vary from clear, honey yellow, orange, brown, red, white and black, as well as rare greens and blues. Overall, there are 256 different colours and shades of amber. The popularity of different colours of amber has changed with changing fashions. In Nero's time, golden amber said to be of a similar hue to his second wife Poppaea's hair was more popular. In the fifteenth century white amber was more popular due to its supposed curative properties. Now, rare green amber seems to be very popular, perhaps due to the rarity of natural green amber, although most green amber on the market is produced by altering the appearance of certain types of amber using heat and pressure to change the colour.

Different shades of natural polished Baltic amber beads.

White amber is opaque due to the large number of minute air bubbles which were trapped within it. There can be almost a million of these minute bubbles in a cubic millimetre of amber. Clearer amber has fewer and larger bubbles. The bubbles act like perforations, making the amber less likely to fracture extensively and making it easier for the sculptor or jeweller to shape. The bubbles can be effectively removed if desired to make the amber appear more transparent by heating it in an oil, such as rapeseed oil or amber oil.

Heating amber can have many different effects on the appearance, clarity and colour. Sun-spangles appear when amber is heated, due to the expansion of larger air bubbles producing a disc-shaped fracture around the bubble as the gas expands with the heat. This effect is quite commonly used in modern jewellery.

Cabochon of rare natural blue amber from the Dominican Republic.

Prehistoric amber

Amber has been used by man for thousands of years. During the late Palaeolithic age (about 50,000–10,000 years ago), much of northern Europe was covered in sparsely vegetated tundra and ice sheets. The people of southern Europe, at this time, were mostly nomadic hunters who travelled many miles following the migrating herds of animals. Very little amber is known to have been used from that time, but occasionally pieces do turn up. Some raw amber that turned up in some caves in what is now the Czech Republic is reported as being over 40,000 years old, from the Middle Palaeolithic. In Spain, a few amber fragments were found associated with human-related debris dated to over 21,000 years ago. The presence of amber here does not, however, indicate that there was an ancient trade link with the Baltic region, but rather that primitive cultures in Spain were using amber found locally.

Amber artefacts such as pendants from Germany and Poland were also being used by people at this time, but were fairly crude in their manufacture, most being raw and unworked pieces of amber. Some pendants were perhaps attached to garments using natural holes in the amber, which would have been easier than trying to drill a hole in the amber, although this was not impossible.

During the Palaeolithic, amber would have been scarce, as much of the amber-bearing sediments and the Baltic Sea would have still been covered by an ice sheet. As amber tends to float in salt water but not fresh water, any glacial meltwaters that filled the area of the present Baltic Sea would have been too fresh to allow the amber to float to shore for quite some time after the ice began to melt. As a result, amber artefacts of this period are very rare.

What exactly were these ancient peoples using amber for? The amber found around this time is clearly not worked in any way, as it occurs as loose fragments, unpolished, uncarved, and unholed. One theory is that they used it as some kind of glue, but this is extremely unlikely, as amber is too difficult to dissolve or melt to be useful in this way. There were certainly plenty of other more appropriate materials around that people could use for this purpose, such as pitch, or living tree resins. It has also been suggested that perhaps it was used for medicinal purposes, or for its fragrance when burnt, as it is today in many cultures. The loose fragments of amber found may be off-cuts from the carving of an amber ornament that has yet to be discovered by archaeologists. Amber may even have served as a currency among some local ancient tribes; but until we have more evidence, amber's early uses will remain a mystery to us.

By the Mesolithic age (about 10,000–7,000 years ago) amber was being fashioned into beads and animal sculptures in northern Europe. Trade between northern tribes became established and artefacts have been found as far west as Britain in sites over 9,000 years old. In Britain, amber was scarce at this time, consisting mainly of irregular lumps. A waterlogged site at Star Carr in North Yorkshire did reveal, however, several irregularly fashioned beads indicating a greater degree of sophistication. Perhaps a reason why there is so little Mesolithic, or Palaeolithic, amber is because other artefacts are found scattered on the surface rather than buried. This environment, so near to the surface, is not conducive to the preservation of amber, where it would dry out and oxidise, eventually crumbling into an unrecognisable dust.

After the disappearance of the ice sheets about 8,000 years ago, it took a while before the North Sea area became inundated by salt water from the Atlantic Ocean. British trade with the amber-producing regions was possible while the land bridge or ice existed, but once this was gone, Britain was isolated from the rest of Europe and mineral industries became very much a local affair in Britain. In Denmark though, the amber industry was thriving and animal carvings and elaborate pendants, which reflected the culture, were being produced. Bears, elk, water birds and humans were all scratched on the surface of amber. Some amber pieces were fashioned into the shapes of tools and buried in graves. Perhaps this indicated the 'profession' of the dead person, or perhaps it was used as a talisman to protect the wearer from harm in the next life?

During the late Mesolithic (8,000–6,000 years ago) amber was more abundantly used by the people of southern Scandinavia. By the early Neolithic (about 6,000–4,500 years ago), the western Baltic regions around Denmark began to use amber in abundance. Grave goods and hoards frequently contain amber items here, but it is not until later in the Neolithic period that amber really takes off in the eastern Baltic regions. During the Middle Neolithic, there is evidence of trade between the western and eastern Baltic regions, as similarly ornamented pieces of amber are found in a number of archaeological sites. By the Late

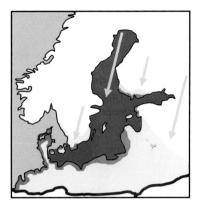

Maximum extent of the glaciation with general ice flow direction and the present day distribution of amber in orange (more abundant) and yellow (less abundant).

5,000 year old amulet from Sweden of a male head.

start to appear in amber ornaments across northern Europe from the Atlantic to the eastern Baltic. The finds of amber in archaeological sites may increase from the Neolithic to the Middle Bronze Age (about 5,000 until about 3,400 years ago), but the pattern of usage stays the same. It appears that the increase in finds relates directly to the growth in the industry and therefore the amount of amber in circulation rather than any changes in local customs.

In Britain there remains a paucity of amber goods from archaeological sites, but Ireland appears to have developed a thriving trade, especially during the Late Bronze Age. In the Iron Age, Britain does not appear to have been a part of the circulation of amber that predominates throughout central and Mediterranean Europe. The socio-economical development of the Alpine and central Mediterranean states during the Late Bronze Age and Iron Age drew much of the amber south from the Baltic supply. Despite this, amber continued to be an important constituent of grave goods until the end of the Iron Age in the Baltic region.

During Neolithic times (about 6,000 years ago), large quantities of amber beads were being used in Denmark. Many of them were found to contain sun-spangles, which are disc-shaped fractures usually caused by heating the amber. This suggests that people had a good working knowledge of amber, and it is thought they were using large pots to heat amber

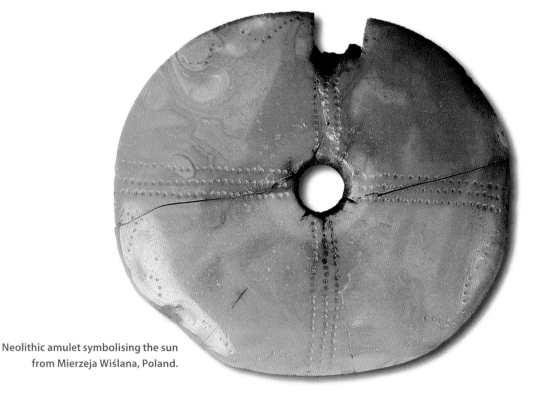

Neolithic amulet symbolising the sun from Mierzeja Wiślana, Poland.

during religious or cultural ceremonies. Thousands of these beads have been found, suggesting that they might have been used for barter and were hence an early form of currency in Denmark. This importance of amber to these early communities may have triggered the later and more widespread trade in amber. It is during this period until about 4,000 years ago that sculptures and beads of all shapes and sizes were produced in amber workshops in Poland, and in towns of other nearby countries, supplied by the Baltic Sea. Elaborate tools of antler, flint, sandstone and animal fur were used to cut, carve, drill and polish the amber. Dotted radial designs on large discs of amber suggest that the Baltic people were possibly sun-worshippers in the latter part of the late Stone Age.

During the later Neolithic, until about 3,700 years ago, amber use in the British Isles was still scarce. It may be that the few amber finds we do have relate to specimens that were washed up on the eastern beaches rather than by trade with the more prolific amber regions. Definite Neolithic finds, however, are known from Scotland at this time. Four irregular beads of amber, associated with jet beads and an axe, were found in a burial mound (over 4,000 years old) at Greenbrae, near Cruden in Aberdeenshire (now in the Arbuthnot Museum). The presence of amber in Neolithic burials has, in the past, been used as evidence for trade with Europe. However, the fact that this amber was irregular in shape, unlike amber being traded elsewhere in Europe, might suggest that it was not traded, but rather collected and worked locally. The amber itself is the exact same Baltic material that was being worked in Europe at that time, but floated across the North Sea to Britain.

The eastern Baltic region of Sambia (the peninsula north of Kaliningrad in Russia) was a major source of amber in prehistory, as it still continues to be today. With rising sea levels over 5,500 years ago, the sediments containing amber would have been washed out, releasing substantial quantities of amber for use by the local populations. The early Neolithic Narva culture of the Lithuanian and Latvian coastline had access to vast quantities of amber between 5,400 and 5,100 years ago. Trade seems to have occurred with the tribes to the north-east as far as the White Sea, which is in present-day Russia and about 2,500 kilometres away! There have been several amber workshops from the Neolithic discovered near Gdańsk that would have helped to supply this trade.

Stone Age peoples were probably aware of the insect and plant inclusions in amber, as these have been found present in some hoards. Amber was valued, not only as beads on a necklace or bracelet, but also

This Neolithic anthropomorphic figure is part of the Juodkranté Treasure from Lithuania now in the collections of the Geological and Palaeontological Institute and Museum of the University of Göttingen, Germany.

3,000 year old Bronze Age amber necklace from Mycenae, Cyprus.

as decorations on clothing and as amulets, perhaps to protect the wearer from physical harm or disease with its 'magical' powers. What must superstitious Stone Age Baltic people have thought when they saw straw or feathers leaping to and sticking to amber when it was rubbed with animal fur? This property where amber holds a static electrical charge was not recorded until the times of the ancient Greeks. Is it possible that this 'magical' property is the reason why it became such a popular adornment in the mid to late Stone Age? Its colour being similar to that of the sun, its unusual light weight, and the ease with which amber can be carved may also have contributed to the widespread use of this unique organic material.

The trade in amber may have existed for over 5,000 years in northern Europe. Amber was one of the principal commodities for barter, and it has been found as far away as central Russia, Norway and Finland at that time. It was not just raw amber that was exchanged, but also the finished product. Amber workshops run by early Baltic tribes produced finished products locally for exchange. Amber provided the Baltic peoples a means to barter for metals that were denied the tribes farther inland, who still had to rely on stone implements.

During the Early Bronze Age, very few amber grave goods are found. This is perhaps because of the practice of cremation; however, we do know that the commerce in amber grew to reach southern Europe. Egyptian blue faience beads have been found in sites of this period in Poland, suggesting that trade extended to Egypt, but not in Baltic amber as far as we know. Some 'amber' that was found in 5,200-year-old Egyptian tombs was, in fact, a much younger sub-fossil resin, or copal. It is also possible that the Assyrians (from an area in present-day Iraq) dealt with the Baltic tribes as far back as 4,000 years ago; however, most of the amber dealing seems to have taken place along European rivers and probably reflects the major commercial routes of the time. The Central European cultures then traded amber from the Baltic region with the early Greeks, who valued it as a luxury commodity for the rich, powerful and educated classes.

In the Iron Age (2,000–1,500 years ago) Gdańsk became the centre of the Baltic amber region and also the centre for the amber trade.

Amber myths

As far back as 2,500 years ago, the ancient Greeks had knowledge of the true nature of amber. The earliest record is that of the Greek philosopher, Thales of Miletos (*c.*635 BC–*c.*543 BC) who described the magnetic properties of amber when rubbed with a silk cloth, noting that amber attracted dust and feathers. At this time, there was a belief amongst the Greeks that amber represented the tears of the Heliades. The story has been related to us by writers such as Herotodus in the fourth century BC and Ovid in the first century BC.

The story is that Phaeton, the son of Helios, rode the Sun Chariot for the Sun God, Apollo. Phaeton was permitted to drive the Sun Chariot across the sky one day, but he lost control of the wild horses that pulled the chariot and came so close to Earth that it was set on fire. This fire was said to be the origin of volcanoes. To save the Earth, Zeus killed Helios' son with a lightening bolt. Phaeton fell from the sky and landed in the Eridanus River (now known as the Po River in northern Italy). Water

The Heliades based a woodcut print for Ovid's *Metamorphoses* in 1563 by Vergilius Solis.

nymphs of the river pulled his body out of the water and buried him on the shore. Phaeton's sisters, the Heliades (also known as the Electrides) went looking for the grave and once they had found it, they vowed to stay with their dead brother, weeping day and night for him. After four months, the Heliades were transformed into poplar trees by the Gods, and their tears into amber. One of the possible origins of the term 'amber' is from the Arabic *haur rumi* which means 'Roman poplar tree'.

THE ROMANS

Although the Romans had a fair idea as to what amber was, they had many superstitions surrounding it. The Emperor Nero used to give amber bracelets to his favourite gladiators, believing it to ensure their survival. He even gave the pet name 'Amber' to his second wife Poppaea because of the golden colour of her hair. Ironically, she was later reputedly beaten to death by Nero.

Certainly, amber as a talisman was much valued by the Romans. According to Pliny the Elder, a small piece of worked amber was said to cost as much as a full-grown male slave. A slave could cost about 600 silver denarii, which is about two years' wages of a legionary soldier. It was also Pliny who suggested that amber was actually the resin of a pine-like tree rather than the poplar tree of Greek legend. He dismisses the myths of the Greeks as nonsense, an intolerable falsehood, and considers it ridiculous that anyone should advance such absurdities. Amber was called *sucinum* (from the Latin *sucus* meaning sap or juice) by the Romans because of its resinous nature, long before Pliny wrote about it in the first century AD.

THE CELTS

Ambres, the Celtic Sun Father, gets his name from the fossil resin. Amber was used extensively by Celtic tribes and they became skilled craftsmen, producing highly prized pieces. It is interesting that the iconic stones of Stonehenge in Wiltshire, England, were once called the Ambers, and the nearby town of Amesbury used to be known as Ambersbury. It may be that it was named after the Romano-British leader resisting the Anglo-Saxon invaders in the fifth century, Aurelianus Ambrosius. Amber has been found closely associated with Cornish stone monuments as well as some barrows (burial mounds) near to Stonehenge. Thus it is possible that there is some link between amber and sun-worship at these sites.

LITHUANIA

In the countries from which amber is derived, there are other legends as to the origins of this precious gem. In Lithuania, the legend of Jūratė and Kastytis is one of the most famous and popular tales. The origins of this story are uncertain, as it was recorded for the first time in the 1800s and has no doubt been influenced by modern romantic stories, as several versions of the plot exist today. The basic story is that the goddess of the sea, Jūratė, lived in a beautiful amber castle under the Baltic Sea where she ruled over all in this domain. One day a young fisherman, Kastytis, while catching a lot of fish, was causing a disturbance. Jūratė decided to punish him, but once she caught sight of the handsome fisherman, she instantly fell in love with him. They were happy together until Perkūnas, the god of thunder, found out. Perkūnas was furious that Jūratė had fallen in love with a mortal. In anger, he sent a bolt of lightning to destroy the goddess' amber castle and kill her mortal lover. Poor Jūratė was then chained to the sea-floor, where she cries tears of amber for her lost lover. The amber that is washed ashore after storms is reported to be either the tears of the mourning Jūratė, or fragments of the destroyed amber castle.

(Photo: Janina Leonavičienė)

Monument to commemorate Jūratė and Kastytis at Palanga, Lithuania.

There is a monument to commemorate Jūratė and Kastytis in the Lithuanian Baltic Sea resort of Palanga. It is located next to a bridge that leads to the setting sun.

Another legend from Lithuania is quite similar. Amberella, the beautiful daughter of a fisherman and his wife, lived on the Baltic shores. One day while swimming, Amberella gets trapped in a whirlpool and is pulled down into the depths of the sea, only to find that she has been captured by the Prince of the Seas to serve as his princess. They live as husband and wife in a submarine amber palace, but she pines to see her parents again. Enraged at her wish to see her parents, the prince mounts foaming white horses, grasps the princess in his arms, and they rise up through the waves in a furious storm. Her distressed parents see their beautiful daughter, adorned in an amber crown and necklace, but held firmly in the prince's grasp. Realising that she will never see her parents again, Amberella tosses lumps of amber that she is holding to her parents on the shore as she sinks back beneath the waves. Her parents grieve as they realise that this is the last time they will see her. It is said that when the Prince of the Seas is angry, storms blow, and Amberella tosses pieces of amber from her prison-palace below to show that she still loves her parents.

POLAND

In the Kurpie region of Poland, there is a myth that has melded with the Old Testament story of Noah and the flood. It shares many similarities with the Greek story of Phaethon and the Heliades as well. When the floodwaters were rising as a punishment to mankind for their sins, people wept in despair. Their tears fell into the water and were turned into amber. Clear amber tears were produced by those who were without sin. This type of 'pure' amber was more frequently used for jewellery. The amber of the sinners was darker and opaque and was mostly used for pipes, or for varnish, rather than jewellery.

Freya wearing the Brisingamen necklace, based on a print in Dahn's (1901) Walhall.

SCANDINAVIA

Freya, the Norse goddess of love, beauty and fertility, had a weakness for beautiful jewels. She was married to the mortal, Odur, and had two lovely daughters. They lived in her palace of Fólkvangr, in the land of Asgard. One day Freya was out for a walk along the border between her kingdom and that of the kingdom of the Black Dwarfs. As she walked she noticed some of the dwarfs making a beautiful necklace. It glistened as golden as the bright sun. Freya asked about this wondrous piece of jewellery and was told this treasure was the Brisingamen, a necklace of great value to the dwarfs. She was desperate to possess such a gem, but was told that all the silver in the world could not buy it. She asked what treasure would buy the necklace, and was told that her love was the price. She was to be 'married' to each of the dwarfs for a day and a night before the Brisingamen could be hers. As if under its spell, Freya was overcome with madness. She forgot Odur and her daughters, and even that she was the Queen of Aesir. No-one knew of the deal for the necklace, except Loki, who always seemed to be around when evil was brewing. After completing the deal, Freya returned to her palace and hid the necklace she had paid so dearly for. Loki went to Odur and told him of what had taken place in the land of the dwarfs. Odur demanded proof of this incredible deal and Loki set out to steal the necklace. He turned himself into a flea and bit Freya while she slept.

This caused her to turn and allowed him to remove the necklace. When Odur saw the necklace, he tossed it aside and left the land of Asgard for a distant land. Freya woke the next morning to find her necklace and husband gone. Weeping, she went to her father, Odin, in Valhalla (which was near to the amber valley of Glaesisvellir), to confess. At the entrance to Valhalla was an amber grove with trees that dripped beads of amber. Although Odin forgave his daughter, he demanded penance. He took the Brisingamen from Loki and commanded Freya to wear the necklace for eternity, wandering the world in search of her lost love, Odur. As she wanders, she continues to weep — her tears that land on soil turn to gold, and those that fall in the sea turn to amber.

Gdańsk and the amber route

The discovery of worked (carved or polished) amber in places such as central Russia, far from its source in the Baltic region, suggests there was an active trade in amber from the late Stone Age, about 5,000 years ago, onwards. Amber was freely bartered in ancient Central Europe and the Mediterranean regions in its raw state, or as worked pieces following the western and central trade routes.

Amber beads and carvings from the late Stone Age have been found in just about every European country. As the Copper and Bronze Ages moved through into the Iron Age, amber really started to gain importance as a commodity. Amber is often found associated with copper and bronze artefacts in countries that have no native copper, which must have arrived there through commerce. This indicates the high value these early communities placed on amber as a trade item. It is perhaps interesting to note that copper and bronze have more practical uses for weapons or agricultural tools, whereas amber was exchanged more for its spiritual or aesthetic value.

In the Iron Age the Aisti tribe of the area around Lithuania traded amber along the Baltic Coast from Klaipeda to the Elbe River. The people who traded with the Aisti tribe were using the rivers as trading highways by which amber was transported long distances from source across Europe. Amber was traded south along the Elbe River as far as the Czech Republic, a distance of over 1,000 km. The route then turned east at Passau in southern Germany and followed the Danube to Linz in Upper Austria. From here the route could access Italy and southern Europe.

There appear to be many branches of the amber route. Some head west to the Rhine, others go east along the Dnieper River to the Black Sea. What is now Gdańsk, in northern Poland, would have been as central to the trade in amber as it is now. Amber gathered from the Baltic coast east and west of the Vistula River would have supplied the central route south to the Mediterranean.

Based on the discovery of amber in some Mycenaean tombs on the Island of Crete, trade in amber to the Mediterranean may have begun as

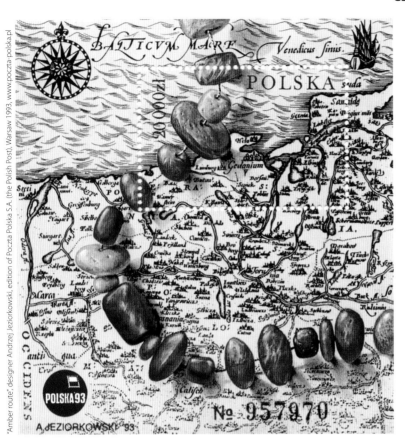

"Amber route", designer Andrzej Jeziorkowski; edition of Poczta Polska S.A. (the Polish Post), Warsaw 1993, www.poczta-polska.pl

Postage stamp commemorating the Amber Route (1993).

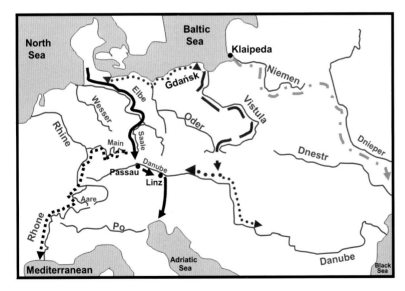

Major rivers that formed part of the amber route from the Late Stone Age to the Middle Ages.

Black = western route;
Red = central route;
Green = eastern route;
Purple = trade along Baltic coast.

early as 3,600 years ago. It is thought that the route for this amber went south along the Vistula River in Poland and then followed the Danube, eventually arriving in the Mediterranean via the Black Sea or the Adriatic

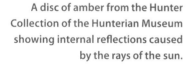
Two zloty coin from Poland commemorating the Amber Route in 2001 showing the route south using the Vistula River during the 1st and 2nd centuries AD.

Sea. The Mycenaean ruling classes in Crete considered amber a luxury and transformed it into objects of adornment such as pendants, amulets, and beads. Quite often, amber is found associated with gold objects, indicating its high status and value.

While all this trade in amber was advancing the cultures of the Baltic coast by allowing them to trade for metals not naturally available to them, it was not benefiting the tribes further inland, many of whom remained in a virtual Stone Age. They were still ploughing their fields with stone tools whilst their richer neighbours in the amber regions to the north had bronze tools. Access to valuable resources such as amber was therefore very important in advancing the culture at this time.

Gdańsk was the centre of the amber trade in the early Iron Age (about 3,000 years ago), the Vistula River being the main artery for the trade south to the Adriatic Sea somewhere near Trieste. We know this route was particularly well used in the first and second centuries AD due to the large numbers of Roman coins from this period found along the route. Not all the amber moving along the route was worked before it left Gdańsk. Several workshops and stores for amber of this period have been found near Wroclaw and Warsaw.

Commerce with the Assyrians in what is now Iraq began as early as 2,900 years ago. Excavations on the banks of the Tigris River revealed amber of Baltic origin. The route that was taken for this amber to get

A disc of amber from the Hunter Collection of the Hunterian Museum showing internal reflections caused by the rays of the sun.

there is unknown, but it is possible that it was traded through Russia, or to the Mediterranean for exchange with the Phoenicians.

The Phoenicians, like many of the European cultures at this time, worshipped many gods, with the Sun God central to their beliefs. Sun-worship practices also played a part in Greek and Roman rituals at this time. Amber may appear to capture the rays of the sun as it reflects light internally, bouncing off flaws and cracks to produce a range of different colours of yellow, orange and white. It is therefore not surprising that amber became so popular and valued amongst these groups. Living along the Mediterranean coast, the Phoenicians were adept sailors and the master businessmen of the ancient Mediterranean cultures. They probably were the 'middle men' of their day, trading their amber through the port of Marseille at the mouth of the Rhone River in southern France, which was at the end of one of the main amber routes direct from the Baltic region. The people who lived there and in northern Italy were called the Ligurians. It is possible that the Phoenicians traded amber with these peoples, as one of the early names for amber was *Ligurian*. The Ligurians presumably then traded with other merchants who came along the various amber trade routes.

Sketch of a Roman carved amber head.

Some Greek and Roman writers suggested that amber had been dug up in Liguria. In fact, as we have read, the Greek myth of Phaethon, the son of the Helios, had him fall into the Eridanus River, which is now known as the Po River in northern Italy. This would have been in the lands of the Ligurians. The true origins of the amber would have remained a secret known only to the traders from the north.

As the Phoenicians were a seafaring culture, they were able to conduct their dealings in amber all around the Mediterranean Sea. The Phoenicians also travelled out of the Mediterranean, where they traded with the British in Cornwall for tin. If they were travelling this far, it is also possible that they were trading directly with the amber producers in the Baltic, cutting out the Ligurian 'middle men'. The Phoenicians also invented stories that the Atlantic was a muddy, impassable sea, full of weird and wonderfully dangerous beasts ('Here be dragons!') most likely to prevent others from following on their valuable trade routes. The Phoenicians were invaded and subjugated by Cyrus the Great in 539 BC, and the strategic Phoenician port of Tyre was later taken by Alexander the Great in 333 BC, ending nearly 700 years of seafaring trade by the Phoenicians. This effectively closed one of the major amber trading routes of the Mediterranean.

The early Greeks continued trading in amber, partly as it had an important connection to their beliefs. Their early writers record the story of the death of Phaethon and his sisters, the Heliades, who cried tears of amber. Interestingly, this story is not too far from the truth of the origin of amber, as the Heliades changed into poplar trees before they cried the tears of amber. This suggests that people were already aware that the 'tears' were made from the resin of trees, although perhaps not the poplar tree.

The trade in amber became less popular as fashions began to change. Although the Greek colonies maintained a strong amber connection, the Greeks themselves were using it less and less. The traders from the Baltic coast did not travel so much with their wares and were content to trade with the neighbouring Germanic tribes east of the Vistula River. Amber did get through to Rome and was reported by Tacitus and Pliny the Elder in the first century AD.

Tacitus, evidently unaware of the history of amber working and trade and the beliefs of the Baltic tribes, wrote:

> ...(the Aisti tribe) even search the sea, and of all the rest are the only people who gather amber. They call it *Glesum*, and find it amongst the shallows and upon the very shore. But, according to the ordinary incuriosity and ignorance of Barbarians, they have neither learnt, nor do they inquire, what is its nature, or from what cause it is produced. In truth it lay long neglected amongst the other gross discharges of the sea; till from our luxury, it gained a name and value. To themselves it is of no use: they gather it rough, they expose it in pieces coarse and unpolished, and for it receive a price with wonder.

There is a certain amount of arrogance in his statement that probably reflects Roman attitudes towards northern tribes at the time: that Romans have the capacity to understand the true worth of amber and that 'barbarians' do not. The truth is that those who collected amber from the coast stood to benefit more by trading it than keeping it, and advanced their culture by doing so. However, Pliny did provide a very knowledgeable account of the origin of amber, and was well aware that amber 'is formed from a liquid that exudes from the inside of a type of pine tree — just as gum in a cherry tree.' He dismisses amber as 'exclusively an adornment for women. Not even luxury has been able to invent a reason for its use.' He used this rational understanding of amber to belittle the Greek philosophers. In fact he states that his description of amber provides him with the 'opportunity for exposing the false accounts of the Greeks', thus indicating the superiority of the Roman intellect.

Roman demand for amber was high in the mid-first century AD. During the reign of Nero, the superintendant of the gladiatorial games, Julianus, sent a Roman knight of the Equestrian Order to acquire amber in order to decorate all the furnishings of the gladiatorial show. He returned to Rome with a large amount of amber that was used for this purpose, the largest piece he brought back being about 6 kg in weight. It is unknown which route he took, or where he ended up on the Baltic coast, but according to Pliny, 'He travelled the trade-routes and the coasts, and brought back such a large amount that the nets deployed to keep the wild beasts off the parapet of the amphitheatre were knotted with pieces of amber.' It is likely that he followed the better-known ancient route north along the Vistula River to the area of Gdańsk.

Although the Romans were ostensibly correct in their interpretation of what amber was, they were unaware that the trees they describe as producing the amber had been gone by many millions of years. The description by Tacitus was taken as fact by the sixth century Roman statesman Cassiodorus who, whilst in the employ of King Theodoric of Italy and the Ostrogoths, wrote the following to the Aisti tribe in response to their gift of amber:

> You say that you gather this lightest of all substances from the shores of the ocean, but how it comes thither, you know not. But, as an author named Cornelius [Tacitus] informs us, it is gathered in the innermost islands of the ocean, being formed originally of the juice of a tree (whence its name sucinum), and gradually hardened by the heat of the sun...We have thought it better to point this out to you lest you should imagine that your supposed secrets have escaped our knowledge.

There is no record of the Aisti tribe's response to the allegation that they were less than forthcoming with the truth, but it is quite certain that the Aisti genuinely had no knowledge of the true origins of amber. King Theodoric became quite interested in Baltic amber and sent several expeditions to find some of the 'Gold of the North'. His expeditions were so successful that he was even able to secure a nugget of amber weighing about 3.5 kg.

In the northern parts of the Baltic region, the Vikings dominated the sea trade routes from the eighth to the tenth centuries. Amber was an important commodity for the Vikings and was used for jewellery,

gaming pieces and religious artefacts. Amber grave goods of this period are quite commonly found associated with Viking burials, including the British hoard from the Knowe of Moan in Orkney. The Vikings were not alone in the Baltic Sea for long, though, as in the eleventh century AD, the Curonian pirates from Lithuania and Latvia grew in power and challenged the Vikings, frequently plundering the Scandinavian coast. Both the Vikings and the Curonians became rich, partly from their spoils and from trade in amber.

It was during the tenth century AD that the region that is now Poland was converted from their panentheistic and animistic pagan beliefs to Christianity. The use of amber was in decline at this time, perhaps as a result of a change in fashion or the conversion from pagan religions, as the wearing of amulets and beads may not have been very popular with the early Christian missionaries. The few amber goods from that period continue to consist of amber beads, presumably a persistence of the superstitious beliefs of the protective or curative power of amber. Amber crosses become more common amongst the converted populations, but the practice of Christians burying their dead without grave goods makes it difficult to be certain what effect Christianity had on the amber trade.

Amber and the Teutonic Knights

AMBER TRADE AFTER THE FALL OF THE ROMAN EMPIRE

The collapse of the western part of the Roman Empire in the fifth century AD led to the collapse of the amber trading routes between the Baltic states and the rest of Europe. The amber trade route persisted towards the east from Gdańsk along the Vistula River and the Dnestr River to the ports of the Black Sea. There was a mass migration of peoples away from the Baltic region towards the east between about AD 400 and AD 700 as interest in amber waned. Whether the trade in amber played an important or subsidiary roll is not clear. It was not until the early Middle Ages that the Slavic peoples of eastern Central Europe once again saw amber as a primary material for jewellery and ornaments. At this time, a 'class'-based society was being established in the Baltic region with feudal states governed by monarchs and their aristocracy. During the tenth century, small amber crosses were being fashioned for the first time. By the time of the Scandinavian Baltic crusades, amber workshops were

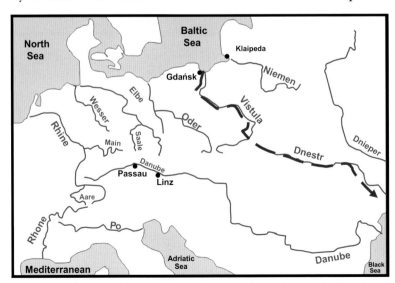

Major rivers that formed part of the Amber Route (red) after the decline of the western Roman Empire.

again thriving in the Baltic coastal towns such as Gdańsk. Amber used to belong to anyone who found it, but with the advent of the feudal system in the Baltic region, the Dukes of Pomerania had laid claim to all the amber found on the coast of northern Germany in the west and as far as Gdańsk in the east.

THE CRUSADES

Before the Teutonic Knights arrived in the Baltic region, Pope Alexander III had authorised crusades against the pagan peoples of the eastern Baltic in the 1170s. The tribes fought back, but some surrendered, accepting Christianity, only to revert to their pagan religion once the invaders had left. Nyklot, leader of the Obotrites Slavs from west of Pomerania (now part of northern Germany) did just that and avoided any loss of territory. Not all were so lucky, though, as the Pomeranian Slavs surrendered to King Canute VI of Denmark in 1185 after having 35 ships captured by just nine Danish ships in thick fog. In 1190, the Military Order of the Teutonic Knights was set up in Palestine, consisting of mainly German knights. It was one of the three main Military Orders of the crusades, on a par with the Templars and the Hospitallers. On their return from the Holy Land, the Teutonic Knights were employed to suppress rebellion in various European states, and were invited to help establish control in the Transylvanian state of Burzenland by King Andrew of Hungary. The Teutonic Knights had ambitions to become more powerful and wealthy like the Knights Templar, and began to erect stone fortifications in place

The tribes of the Baltic region and the crusades by the Danish and Swedish armies in the 13th century.

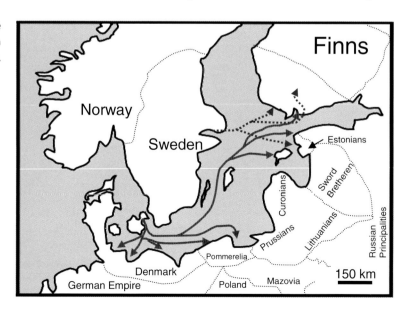

Above: Map of present day Poland with important cities and teutonic strongholds. *Left:* Drawing based on a copper engraving of amber fishermen and gallows by Wagner (1774).

of the wooden ones they were permitted to build. King Andrew was then pressured by his own barons into removing the Teutonic Knights from the region in 1225.

THE TEUTONIC KNIGHTS

In 1226, the German Emperor Frederick II granted Prussia to the Teutonic Knights. The Teutonic Knights had learned from their mistakes and, when asked to help Duke Conrad of Mazovia against the pagan Prussians, made sure that they set up an independent centre of operations around the settlement of what is now Chelmno in Poland, just over 100 km south of Gdańsk. Control of the Baltic amber trade was about to change yet again with the arrival of the Teutonic Knights. These crusading knights set up home along the Baltic Coast, imposing stringent regulations on any commercial enterprise in the region. From this presence in the Baltic, the Teutonic Order soon dominated the region.

From Prussia, the Teutonic Knights fought their way through the Baltic tribes, taking part of western Lithuania and defeating the western Prussian tribes. The Baltic Crusades were financed mostly by the northern Germanic Christian monarchies, who were perhaps less interested in the journey to Jerusalem. Run by a Grand Master, their Order eventually chose Malbork to be their administrative centre. The castle of Malbork, previously one of their strongholds, was, as a consequence, converted into a citadel and residence for the Grand Master. The Teutonic Knights ruled the trade in amber with a hanging noose. Anyone found collecting amber and not in possession of a licence was executed. Licences, however, were difficult to come by, and in the early days of the Teutonic Order, only the Bishop of Sambia (the peninsula to the north-west of Kalningrad) and the fishermen of Gdańsk were allowed to fish for amber. This, however, did not last long. As the price of amber increased, the licences were revoked, and anyone found collecting amber, or even in possession of raw amber, was hanged from the nearest tree. A beachmaster, who oversaw the collection of amber for the Order, would have carried out these executions. This remained the case into the fifteenth century.

Amber collected by the Teutonic Order was destined for the rosary bead trade and was stored in warehouses away from the region to prevent illegal workshops from being established near to the source of the amber. The amber craftsmen further protected themselves from illegal workshops by setting up guilds. The first amber guild was set up in 1302 in Bruges, in what is now Belgium. This was soon followed by the craftsmen of Lübeck, in northern Germany, who set up another guild in about 1310. These guilds supplied the entire Christian church with amber rosaries during the early 1300s. The Baltic region was a place of turmoil during much of the Middle Ages as different states fought to consolidate and forcibly expand their boundaries, some using conversion to Christianity as their motive in order to obtain papal support. During the fourteenth century, the Baltic region became the favoured area for crusading. The pagan Lithuanians were especially targeted to provide the crusading armies with valuable fighting experience. The Papal authorities afforded the Baltic crusades the same tax incentives that the Holy Land crusaders received, and this is why the Baltic Crusades were particularly popular with the Germanic monarchs who sought to expand their borders.

THE TEUTONIC KNIGHTS IN MALBORK

The Teutonic Knights took Gdańsk, the central city of the amber trade, in 1308 and moved their headquarters to Europe's most powerful fort at Malbork, about 40 km southeast of Gdańsk, in 1309. Amber, much valued by the Order, was kept in the treasury along with the silver and gold at Malbork Castle. At the beginning of the 1400s, the rule of the Teutonic Knights was being challenged, and in 1410, the Poles and Lithuanians defeated the Teutonic Order at the battle of Grunwald. The Teutonic Order never regained its former dominance in the Baltic region after this defeat. By the late 1400s amber guilds that were previously outlawed by the Teutonic Order were being set up in various Pomeranian towns such as Kolobrzeg, Koszalin and Slupsk in what is now Poland. The Teutonic Order still wielded some power and took the town of Gdańsk in 1466, placing it under the sovereignty of King Casimir IV of Poland. However, in 1477 the Order was not able to prevent an amber guild being formed in Gdańsk, indicating how strong the guilds had become. King Casimir IV made amber free again, granting the people of Gdańsk the right to control the Baltic shores of Gdańsk and Pomerania.

During the early part of the 1500s, the Order considered Poland its enemy. As a political move it appointed a Grand Marshall from the Polish royal family. Grand Marshall Albert, however, resigned, became a Lutheran Protestant and took with him the rights to the amber monopoly. The conversion to Lutheranism by the German principalities reduced the need for rosaries and, by implication, amber. Albert, now

Statues of past Grand Masters of the Teutonic Order at Malbork Castle.

West side of Malbork Castle.
(Photo: Lech Okonski)

Duke Albert of Prussia, was having difficulty keeping the amber trade profitable, and asked court physicians to investigate the medicinal properties of amber. In addition, he encouraged the use of amber in other works of art, combining it with ivory, tortoise shell, and gemstones, in order to maintain a lucrative market for amber.

With the continued reduction in the rosary bead business, Duke Albert eventually signed over his trade to an agent, Paul Jaski, in Gdańsk. The trade in amber goods moved east to provide beads for Islamic and Buddhist countries as well as elaborate statues for the temples of the Middle East. The guild grew wealthy despite the many disputes over trade restrictions it had with the agent.

Malbork Castle collections

Due to the unique preservation needs of amber works of art, very few really fine historical pieces have been preserved in museums across Europe. As a result it has been extremely difficult for museums to acquire collections that truly represent the diversity of art that was created by amber craftsmen from the Middle Ages through to the twentieth century. Malbork Castle Museum in Poland, which was established in 1961, is one of the few museums that have been successful in acquiring an extremely large collection of high quality amber art pieces that reflect the changes in the craft over time.

The concept of the Castle Museum in Malbork was to create a historically representative collection of amber craft work; a task that proved to be incredibly difficult. At the time that the Castle Museum in Malbork was first established, only bulk-produced souvenir products from the Factory of Amber Products in Gdańsk, and some minor works from the Cepelia traditional craft shops, were available.

The idea of having a major amber collection at the Teutonic castle at Malbork is no accident, as the surrounding area and castle have been linked with amber since earliest times. About 5,000 years ago, the world's largest seasonal amber workshops were situated in the Vistula delta area where the people of the Neolithic Rzucewska culture used to trade amber. Later, in Roman times, Malbork was close to where the amber route turned east towards Sambia. During this period, settlements rapidly developed on the right bank of the Vistula River and the Nogat distributary, especially in the areas of present-day Gościszewo and Wielbark, where many Roman artefacts have been found during archaeological excavations carried out by the Castle Museum. In the tenth century, rings, gaming tokens, beads, necklaces and pendants became the specialities of the Gdańsk craftsmen. In the thirteenth century, Malbork became the centre of power of the Teutonic Knights, who controlled the rights to the raw amber in the Baltic region for over two centuries. Malbork Castle was therefore the perfect location to tell the story of the development of amber artefacts in the Baltic region.

Map of the area around the Vistula delta.

Above: Neolithic beads of the Rzucewska culture (2,100–1,750 years BC); *Below:* Roman necklace of amber and ceramic beads from a burial at Weilbark (1st–4th century AD).

9th century amber beads and conical gaming pieces from the Truso settlement near Elbląg (present-day Janów Pomorski).

The collections were substantially enhanced during the 1970s, with the help of Franciszek Studziński of Paris, by the acquisition of valuable pieces, and also the transfer to Malbork of a dozen or so antiques from the National Museum in Warsaw. Examples of amber craftsmanship continue to be acquired today through antique auction catalogues, private purchases, and by donations from the public. It was a long and arduous process, since old amber craft pieces are very rare and difficult to find. Despite these difficulties Malbork Castle Museum has managed to acquire over 2,000 amber exhibits represented by material from the Neolithic Rzucewska culture to the present day, concentrating mainly on artistic artefacts. In order to understand and display amber in all its forms, the collection grew to include significant pieces of raw amber, as well as some with inclusions of insects and plants.

An amber ship by Wiesław Książek (1971).

For nearly 50 years the exhibition at Malbork Castle has welcomed millions of tourists from all over the world, helping them to discover and appreciate the value of Baltic amber in the development of traditions and art in Poland. For many people today, amber may be a relatively unknown material, almost exotic, but it has had an important role from prehistoric times in the economic and artistic development of Europe. To promote amber craftsmanship to a wider audience, the collections at Malbork have been shown in many prestigious museums around the world, including Europe, the United States, Japan, and in 2010, Scotland.

Other museums with significant collections include Catherine's Palace in Pushkin, the Green Vault in Dresden, the Museum of Art History in Vienna, the Museum of Decorative Art in Berlin, the Victoria and Albert Museum in London, and in Skokloster Castle collections near Uppsala. The collections of Malbork Castle Museum, however, provide a unique insight into the development of amber craftsmanship and changing amber fashions throughout time, especially from the sixteenth century onwards.

FURTHER DEVELOPMENT OF AMBER CRAFT

Sixteenth century

Many of the pieces from the sixteenth century in collections from around Europe are recorded as having come from the workshops of Kaliningrad in Russia. Most of the pieces from this period were of a small size, being limited by the size of the raw material, and the fact that the craftsmen had not yet acquired the knowledge and skill to meld small pieces of amber together.

Amber necklace with facetted beads of Sybille Dorothy.

Although the largest part of the collection is from the seventeenth century onwards, there are two important and extremely rare pieces of jewellery from the sixteenth century. One is a series of 10 large faceted beads from the necklace of Duchess Sybille Dorothy Hohenzollern (1590–1625), which originally consisted of 23 beads and until the Second World War, was in the collection of the Silesian Museum of Applied Art and Antiquities in Wrocław, Poland. The other is a necklace of 49 cylindrical beads of translucent amber decorated with garlands in low relief that were used as an adornment for a man of wealth and position in Gdańsk.

Amber processing in Gdańsk at the beginning of the sixteenth century still focused on religious themes, but the situation was soon to change as the Reformation movement in the second quarter of the century reached the outskirts of Gdańsk. This movement greatly influenced the diversity

Cylindrical decorated amber beads for a man's outfit made in Gdańsk about 1600.

of craft manufacture, with local craftsmen now finding it possible to use amber for a much wider range of purposes. This is highlighted by a correspondence of the Papal envoy, Cardinal Francesco Giovanni Commendone (1563), 'Amber is now used to manufacture cases, spoons, little vases, even birds' cages; everything that due to its fragility is there more for the eye, than practical purposes'.

Although the diversity of products had increased, the limited access to raw amber in the Bay of Gdańsk resulted in these products becoming luxury items, and expensive ones. The number of amber workshops had been limited to 40 by the Gdańsk City Council in 1549, due to the difficulty in obtaining the raw material. One means of solving this problem was to import amber from Prussia, but for nearly a hundred years the Jaski family had held a monopoly on the trade of Amber not only in Gdańsk, but also along other Baltic Coastal towns. The Jaski company enforced restrictive operating conditions on the amber craftsmen, and it was not until the death of Paul Jaski at the end of the sixteenth century that this situation began to change for the better, although the company maintained the monopoly on Prussian amber until 1642.

Amber craftsmen from Gdańsk quickly gained an esteemed reputation among the upper classes of the Commonwealth of Poland, including the royal court. Exceptional, unique, and thus expensive and sought-after goods of the local masters served as decorations of both attire and interiors in the sixteenth and seventeenth centuries. Kings also ordered them as diplomatic gifts, and were themselves keen collectors.

Seventeenth century

Towards the end of the sixteenth century, a new method of working amber was developed that allowed larger, more complicated pieces to be produced. Artists mastered the art of carving small plates from the amber, which they sculpted into small reliefs, or engraved from beneath. These plaques were then joined together using glue or silver frames. This technique is credited to the Kaliningrad amber master Georg Schreiber, and is the way the first amber caskets were made. Amber caskets were mainly used for storing jewellery and important documents, and the workshops of Gdańsk were especially renowned for these. In the Malbork collection there are five of these fine caskets, the earliest of which dates to the early seventeenth century. It is a large sarcophagus-like casket, having no wooden structure, and made entirely from multi-coloured layers of amber cut out in the form of various geometric shapes. It is distinguished by twisted pillars with Corinthian capitals on the sides of the casket.

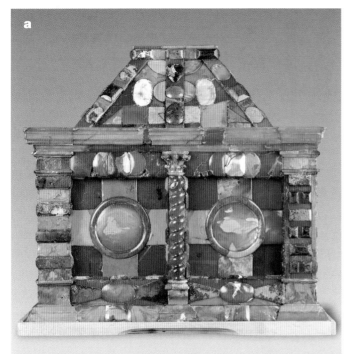

Jewellery boxes were the most popular products of the local master craftsmen. The transparency of some of the amber was used to its full advantage by carving miniature scenes such as landscapes or plant motifs on the reverse side of thin plaques of amber. These were then cut into various shapes and inlayed onto the boxes. In the mid-seventeenth century, wooden frameworks began to be used for the construction of amber products, and the effect of clarity was substituted by strong colour contrasts in amber. Placing sheets of gold leaf under the engraved amber plaques gave the engravings more life, and added luminosity and gleam to the whole piece (the eglomisée technique). The major achievement

Caskets from the 17th century in the Malbork Castle Museum collections: a) side view of casket with twisted pillar and Corinthian capital (early 17th century); b) casket from the workshop of Michel Redlin (about 1680) with; c) detail of the medallion from the casket floor with a scene made using the eglomisée technique; d) casket of Stanisław Leszczyński in side view (about 1700).

of the amber craftsmen from Gdańsk was the skilled way in which they chose the colours and contrasts of the different types of amber, to produce a beautiful mosaic colour effect. By comparison, those from Kaliningrad appear very monochromatic. The different colours of the raw material were also used to differentiate between the individual spaces on gaming boards, and game pieces, such as in chess.

Another casket in the Malbork collection is traditionally linked to King Stanisław Leszczyński. It is the only one to have a wooden skeleton framework. It is pasted with multicoloured amber and decorated with medallions made with the use of eglomisée technique featuring stylised plant and floral motives.

Ivory was often used in combination with amber. In early pieces this was in the form of plaques or medallions with carved flower motifs or mythological/biblical scenes and can be found on amber jewellery cases, candlestick holders, or family altars made in Gdańsk. Amber craftsmen from Gdańsk appear to have been much keener on the use of ivory than craftsmen from other amber producing centres, such as their main rivals in Kaliningrad.

The artistic processing of amber, and the characteristic features of the flourishing crafts of Gdańsk, were recorded in the diary of Papal auditor Giacomo Fantuzzi of Ravenna during his stay in Gdańsk in 1652: 'Amber is elegantly processed in Gdańsk, both the yellow and white kinds, the latter being of much greater worth, as it is used for sculpting small figurines, which are later embedded in yellow amber; moreover they are so finely made that they appear real, and it is with great interest that one observes how they sculpt them with nothing else but the tip of a simple knife.' Historical sources such as these also provide us with information about some of the amber pieces that have not survived to the present day.

Polish monarchs often visited Gdańsk to obtain gifts of amber for other members of European royalty or members of their retinue. According to a report of the visit of King John Casimir and his wife in 1651, the Abbot of Oliwa, Aleksander Kęsowski, presented the king with an amber clock: 'a costly, tall piece, for which he paid three thousand zlotys in Gdańsk'. At that time, it was not only the Polish monarchs who ordered amber products from Gdańsk. Duke Frederick Wilhelm continued the tradition of presenting diplomatic gifts initiated by the Grand Masters of the Teutonic Order. Though he made use of the services of amber craftsmen from Kaliningrad more often than those from Gdańsk, it was from Gdańsk that he ordered a throne for Emperor Leopold I in 1676;

three gift pieces for Versailles in 1679; a table for William III of Orange, as well as a huge mirror frame. This same mirror frame, made by Nicolaus Turau, can be found today at the Museum of Decorative Arts in Berlin.

Other artists from Gdańsk, especially those operating in the last quarter of the seventeenth century, also made a name for themselves as the finest in Europe. Michel Redlin was a supplier of fine amber pieces to the Russian and Swedish courts, and Christoph Maucher, an artisan working outside the guild, became the most famous amber craftsman in Gdańsk, supplying his wares to various royal courts all over Europe.

Maucher came to Gdańsk from Southern Germany around 1670, and continued working there until continuing protests from the city's

Late 17th century baroque chest made by Cristoph Maucher using different kinds of amber with elaborate sculptural decoration and four legs connected by a full relief sculpture and a central figure of Neptune.

Detail of the figure of Neptune beneath the Cristoph Maucher baroque chest.

guilds forced him to leave in 1705. Familiar with processing ivory in the workshop of his father, who was specialised in decorating the handles of muskets, he quickly mastered the art of working with amber, which required fairly similar skills. Maucher was probably trained by Gdańsk master Nicolaus Turau, and with the assistance of Gottfried Wolfram from Turau's workshop, the first man to work on the Amber Room, Turau and Maucher made the throne chair for Emperor Leopold I.

The most beautiful casket from the Malbork collection originates from the workshop of Christoph Maucher. It is a wonderful piece, characterised by rich baroque sculptural decoration of a maritime theme, and the goddess Venus. Four legs linked by mythical horses driven by cupids, and a chariot featuring Neptune in the centre, support the casket body. The sides are engraved with acanthus runners, faceted cabochons, and seashell reliefs. Central to all of this is a statue of Venus, the goddess of love, in a seashell-shaped bay, surrounded by reliefs featuring animals, garlands with fruits, and vases with plants made from white amber. The lid is topped with a figurative group representing three goddesses from the scene of the trial of Paris. In the interior of the lid there is a large oval medallion presenting Venus and Mars resting under a tree and a cupid that accompanies them. Set in the bottom of the casket are four medallions featuring scenes from the life of Venus as well as a fifth central one, made using the eglomisée technique, featuring Venus riding in a flying chariot.

Michel Redlin's workshop is linked with a casket that has a drawer for writing instruments (inkstand and sand-box), with lid adorned with a statue of Ceres, the goddess of fertility. In the casket body, there are oval medallions with maritime scenes manufactured using the eglomisée technique, as well as transparent plates featuring floral and arabesque ornamentation (see picture **d**, p. 41).

The small amber boxes for snuff or perfumes are also very good examples of the fine work being produced at this time. The lids of these boxes were adorned with mythological scenes such as Diane bathing or Narcissus admiring his reflection in the spring. On a smallish bottle linked with the Christoph Maucher workshop, naked cherubs entangled in acanthus leaves have been finely sculpted.

Phial decorated with naked cherubs among acanthus leaves from Cristoph Maucher's workshop (end 17th century).

Small amber boxes made in Gdańsk at the end of the 17th century. Low relief decoration on the lids, of Narcissus at the spring (above) and Diana bathing (below).

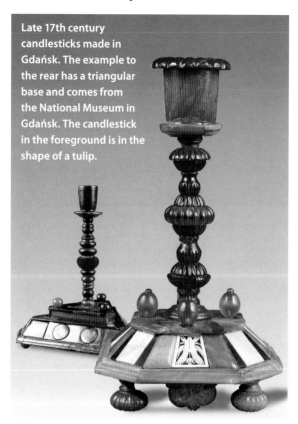

Late 17th century candlesticks made in Gdańsk. The example to the rear has a triangular base and comes from the National Museum in Gdańsk. The candlestick in the foreground is in the shape of a tulip.

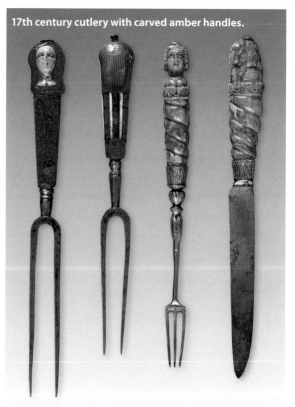

17th century cutlery with carved amber handles.

Around this time carved amber was also used to decorate the handles of cutlery. One piece of particular interest has the inscription: *Ioannes Henricus Herman Rosenthal Anno 1678* carved into transparent amber using the eglomisée technique, providing us with a rare precise date of manufacture. It is very rare that amber pieces feature a date. From the Malbork collection only two other pieces can be dated in this way; one is on a baroque home altar and the other on a nineteenth century clock from Kaliningrad.

Amber candlestick holders were also being made at the end of the seventeenth century, and were often elaborate floral designs. One piece is decorated with plates of mother of pearl and relief ivory plaquettes, into which is set a neck made of a number of flattened amber rings terminating in a candle holder the shape of a stylised tulip flower.

Many amber products at this time were still made to supply a demand for religious pieces. A fine ornate example of this type of work is a large multi-storey home altar, dated to 1687, each consecutive storey being flatter and shorter, to produce a slender, towering shape. A veneer of differently coloured pieces of amber, as well as ivory, applied to the altar provides a multicoloured decorative effect. Ivory was used for the

Rare dated architectonic amber altar on a wooden framework decorated with ivory sculptures and reliefs from Gdańsk (1687). Detail of amber work on right (top) the Crucifixion; (middle) the Last Supper; (bottom) the Nativity with date on lower left.

Late 17th century miniature altar with the figure of Madonna and Child, decorated with ivory medallions depicting scenes from the life of Christ and Mary as well as ivory plaquettes with floral motifs.

Amber reliquary in the form of a monstrance (liturgical vessel). The central recess for the relics would have originally had small doors. The piece is decorated with amber and ivory low reliefs representing scenes from the life of Christ, Saint Helen and other scenes in the top section.

statue of Christ and angel as well as the central relief illustrating the Last Supper, several plaquettes of various scenes from the New and Old Testaments, and balustrades in the lower section. On the top of the altar there is an amber figure of the resurrected Christ.

Another example is distinguished by its large statue of the Virgin Mary with Child above a crescent moon. This little altar is quite an unusual design for its type, as they are usually crowned with a crucifix. The individual storeys of the altar are separated by twisted amber pillars and are decorated with oval and rectangular ivory plaquettes of floral motifs and scenes from the life of Christ.

Eighteenth century

The eighteenth century amber reliquary of the Michel Redlin workshop is one of the finest pieces of amber workmanship in Europe. It has been exquisitely carved in the form of a monstrance (vessel for carrying religious relics) which is richly decorated and set on a slender trunk with an ornamental nodus half way down, and a base that is adorned with the winged heads of angels. Four reclining amber lions at the base, looking too delicate to support its weight, give the impression of the entire structure rising into the air. The amber craftsmen have succeeded in creating an impression of something ethereal, and unreal.

From the end of the Teutonic occupation in 1454 until the beginning of the eighteenth century, amber craft in Gdańsk was among the leading professions, and amber craftsmen were considered one of the most prestigious vocational groups. At the beginning of the eighteenth century, amber craftsmen from Gdańsk were employed in the creation of the most impressive piece of amber art; the Amber Room. The fame gained by this royal amber study did a great deal to promote the specialist craftsmen and artists amongst rich merchants and the nobility in the early part of this century. Amber items were often collected to be included in cabinets of curiosities, which were very fashionable from the end of the seventeenth century into the eighteenth century. These ranged in size from pieces of table-top furniture to entire rooms filled with an eclectic mix of curious objects collected by, or for, the owner. All of this helped to raise the prestige of the amber craft at this time.

The Malbork Museum collection of eighteenth century amber artefacts reflects the less grandiose style of amber craftsmanship that began to appear at this time. The majority of items are small ornamental trinkets and party games, often with an inscribed insightful phrase. An example is a solitaire game board with top divided into twenty squares separated

Early 18th century amber chess pieces.

Solitaire game from the 18th century.

by rosettes of light amber. Between the playing area there are four medallions with engraved scenes and fashionable French inscriptions such as, 'Curiosity is dangerous.' The natural variation in the colour and opacity of amber also made chess a popular product of amber workshops since well before the eighteenth century.

An important piece of eighteenth century amber work surprisingly turned up in Scotland. How it got there is a bit of a mystery, but it once belonged to the last king of Poland, Stanisław August Poniatowski, who was forced to abdicate in 1795, move to St Petersburg, and later died deeply in debt in 1798. Lady Barbara Carmont of Edinburgh donated this important piece to the collections at Malbork Castle Museum. According to Lady Carmont the cabinet made its way to Italy after the death of the king, where it was purchased by an ancestor of Lord Carmont, who at that time was a treasurer at the Vatican. The cabinet, apart from

its undeniable artistic value, is of immense historical value to Poland, as it contains engraved scenes and inscriptions relating to the most significant events from the life of Stanisław August. Even more important are some personal inscriptions hidden beneath the amber medallions, which express the bitterness felt by the monarch after he was almost kidnapped by the Bar Confederation (an association of Polish nobles) in November 1771. The cabinet is in the shape of a miniature baroque cabinet filled in with drawers. In the central recess is an earlier seventeenth century statue of the Virgin Mary with Child. Perhaps Stanisław August had this piece of furniture made with the intention of placing the amber Madonna in the recess as a votive offering for being rescued.

A great deal of information about the manufacture of amber has been obtained from contemporary documents and designs such as the bill made out by Michel Redlin for the completion of three amber items, ordered by Frederick III as gifts for Russian tsars. The pieces described in the document, a chandelier, a jewellery case and chess set, were most probably destroyed during the 1737 fire of the Moscow Arsenal. Luckily, the artist had attached the drawn designs to the bill, and these very

Miniature baroque amber cabinet of King Stanisław August Poniatowski containing a figure of the Madonna in a recess surrounded by 11 drawers. Made in Gdańsk after 1771. Donated by Lady Barbara Carmont of Edinburgh to the Malbork Castle Museum collections in 1979. Inscriptions in German include statements by King Stanisław such as 'I'll be back', 'Life from death', 'Weak soldiers seek peace', and memories of events such as 'The oath taken by Stanisław Augustus forced by Kosiński in the Bielany Forest on November 3rd 1771'.

Michel Redlin sketches of some of the gifts intended for the Russian Tsars Ivan and Peter Aleksievevich.

precise drawings allow other unsigned works to be identified stylistically as being from the Redlin workshop.

Around the mid-eighteenth century, there was a slow decline in the production of amber art. This was a result of a number of factors such as the availability of the raw material, and less demand for the large and expensive pieces which had previously been favoured by the Gdańsk nobility and royalty alike. The final straw arrived when the city was incorporated into Prussia in 1793 and the city's economic superiority began to crumble due to the Prussian wars of the eighteenth century.

Nineteenth century

The political situation in the nineteenth century resulted in Gdańsk becoming a provincial Prussian town, as Kaliningrad began to play a leading role. On the Sambian peninsula near Kaliningrad, amber extraction and processing started to assume an industrial scale, with the raw material itself being sold in large quantities around the world. The main amber-based products manufactured from this extraction were centred on the tobacco industry, with Vienna workshops becoming particularly

famous for producing amber mouthpieces for pipes and other smoking paraphernalia.

Amber was only used for the mouthpieces, due to the fact that it burns when lit. The central part of the pipe was most often made from meerschaum or silver. The silver was usually adorned with a filigree lace-like decoration, forming stylised plant motifs or occasionally geometrical ones. Meerschaum, a clay mineral also known as sepiolite, which was imported from Turkey, was the most common ornamentation for pipes. Meerschaum was easily worked, resulting in elaborate relief sculpted figures of people and animals as well as carved initials and family crests, as adornments on the pipe stems. The attached amber mouthpieces were usually either bent in an 's' shape, or straight and flattened.

Apart from the accessories for smokers, which were the most common use of amber at that time, amber jewellery products, such as tiny boxes, pendants, necklaces, and brooches, were also being made, sometimes with gold or silver findings.

The Amber Room

The original Amber Room was created in the early 1700s by King Friedrich I of Prussia. Once completed it was gifted by his son, King Friedrich Wilhelm I, to his ally Tsar Peter the Great of the Russian Empire. An incredible 6 tonnes of amber had been used to decorate the walls, which consisted of amber panels backed with gold leaf. During the Second World War, the room was looted by the Nazis, along with many other works of art, and its whereabouts has remained a mystery to the present day. What had been dubbed the 'Eighth Wonder of the World' became one of the greatest mysteries of the twentieth century. There are many improbable theories and unlikely conspiracies about the Amber Room that make it difficult to distinguish fact from fiction. In 1979, the Amber Room was recreated at the Tsarskoye Selo (Tsar's Village), near St Petersburg. Present day amber craftsmen had to rediscover old techniques and develop new ones to shape, work, and colour the amber. The new Amber Room was eventually finished in 2003 for the 300th anniversary of the city of St Petersburg.

The construction of the original Amber Room began in Prussia in 1701 at the Hohenzollern Palace (Stadtpalast) in Berlin on the orders of King Friedrich Wilhelm I of Prussia, who had it created at the request of his second wife, Sophie Charlotte. The room was designed by the royal architect and sculptor from Gdańsk, Andreas Schlüter, who also worked as an architect for the Russian Tsar Peter, and designed the façade of the Royal Chapel in Gdańsk, as well as producing many architectural and sculptural works in Poland for King John III Sobieski of Poland and others. Sadly, most of the architectural work he had produced on state buildings in Berlin was destroyed in the Second World War.

Gottfried Wolfram, a Danish amber master craftsman who came recommended by the Danish King Frederick IV, was afforded the task of constructing the chamber in Berlin. Wolfram and Schlüter were eventually relieved of their duties in 1707 as a result of 'court intrigue'. Amber masters Ernst Schlacht and Gottfried Turau of Gdańsk were brought in to take over. Sophie Charlotte died in 1705 before its completion, but the

Catherine Palace in St Petersburg.

© Herbert Spichtinger/Corbis

work continued on the Amber Room until 1713 when Friedrich I also died and his son Friedrich Wilhelm I succeeded him as King of Prussia. All work on the Amber Room was stopped, as Friedrich Wilhelm I considered it an unnecessary expense. It was catalogued and boxed by Turau and Schlacht. In 1716, Friedrich Wilhelm decided to present Tsar Peter I with the amber to help secure an alliance with Russia, against their common enemy Sweden. Schlüter, then working for the tsar, had previously told him of his unique design.

The crates containing the amber arrived intact in St Petersburg in July 1717. Despite his enthusiasm for this work of art, Tsar Peter was not able to complete the installation of the room before he died in 1725. The Amber Room was not installed until the Tsarina Elizabeth of Russia had it rebuilt in the Winter Palace in St Petersburg, where it was used for official functions between 1746 and 1755. It was then moved to the Tsars' summer residence at Tsarskoye Selo, 25 km to the south-east of St Petersburg at the Catherine Palace. More Baltic amber was sent over from Prussia in order to expand the Amber Room and incorporate some new designs by Bartolomeo Rastrelli, the architect of both the Winter and Catherine Palaces. Apparently, the finished room was dazzlingly lit by 565 candles reflecting warm golden light off the mirrors, gilt, and amber mosaics.

The original Amber Room before it went missing during the Second World War.

Creator: Sergei Mikhailovich Prokudin-Gorskii

The Amber Room remained relatively untouched, except for a few eighteenth century renovations and conservation works, until 1941, when Germany invaded the Soviet Union. The invasion was code-named *Barbarossa* after the twelfth century leader of the third crusade to the Holy Land, Emperor Frederick Barbarossa of the Holy Roman Empire. The curators responsible for safeguarding the Soviet art treasures had trouble removing the panels, as the amber had become brittle over the years. They opted to cover the panels with wallpaper in an attempt to foil the Nazi raiders. Unfortunately, and perhaps not unsurprisingly, this ruse did not fool the Nazis, and they removed the Amber Room in a matter of a few hours. It was then transported to Kaliningrad (then Königsberg) where it was placed partly on display, but mostly in storage at the castle.

In 1945, Hitler gave the order that allowed the movement of cultural goods from Kaliningrad. Eric Koch, a Nazi party leader in Prussia, was in charge of the operation, but nobody knows for certain what happened to the crates that contained the valuable works of art from Kaliningrad. Some say they were transported by train, or buried in old mountain mines, and others that they went onboard the passenger ship the *Wilhelm Gustloff* which was sunk by a Soviet submarine, or perhaps it was even taken to Germany. Kaliningrad itself was heavily bombed by the Royal Air Force and severely damaged by the advancing Soviet Red Army, and hence the Amber Room itself may have been destroyed in these attacks, although the amber was removed to the basement of the castle after the bombing raids of 1944.

Several books have been written on the mystery of what may have happened to the Amber Room panels. One such book written by two investigative journalists, Catherine Scott-Clark and Adrian Levy (2004), suggests that it is likely that the Amber Room was destroyed by the Red Army in 1945. The panels seem to have been packed in crates in Kalinigrad Castle ready for transport to Germany when the Soviets arrived. Soon after the castle was taken, the room in which the works of art were stored was destroyed by fire. Perhaps the Amber Room had been taken by the Soviet troops, or perhaps it was destroyed in the fire started by the Soviets? Certainly, this is highly contentious, and the Russians hotly dispute this account, as an envoy sent by them in 1945 reported the Amber Room missing rather than destroyed. Other theories suggest that the Amber Room was found and hidden by the Soviets in order to blame the Germans for its theft. This theory was compounded by the Soviets not allowing access to the castle ruins by archaeologists and historians, and the ruins being blown up in the 1960s by the army to facilitate the construction of the 'House of Soviets' in Kaliningrad. The Russians today still do not accept that the Soviets were responsible for the 'disappearance' of the Amber Room in any way. Now, after such a long time, it is unlikely that the Amber Room panels would have survived without proper care. Even so, there are still claims by some well-funded treasure hunters that the Amber Room is still out there and only they know where it is.

Although we may never recover the Amber Room, pieces have occasionally turned up, such as a small charred fragment that was discovered by a Soviet investigator. Another piece, allegedly stolen by a German soldier who helped pack the amber panels, turned up for sale in Germany. It was the head of a warrior from a Florentine mosaic that was part of a set of four. However, it provided no clues as to the whereabouts of the rest of the Amber Room. Perhaps one day more evidence will turn up to provide us with at least a glimpse of what really happened to the original Amber Room.

In 1979, a reconstruction of the Amber Room began, based on documents produced by Turau and Schlacht of the original room, and pre-war photographs of the room at the Catherine Palace. It had a difficult beginning due to lack of finance, but the project was rescued in part by Ruhrgas (a German gas company with interests in Russia) after

The new Amber Room.
(© Roland Weihrauch/dpa/Corbis)

the Soviet government withdrew the project's funding in the 1990s. The amber craftsmen had to relearn old carving, moulding, colouring and fitting techniques and develop new ones in order to complete the project by the 300th anniversary of St Petersburg. The total cost of completing the Amber Room is estimated to have been over £7 million.

Finally complete, the new Amber Room was opened by the Russian President Vladimir Putin and German Chancellor Gerhard Schröder at the Tsarskoye Selo on the 31st May 2003.

Amber into the twenty-first century

Changes in Polish art have been heavily influenced by political change in the twentieth century. The two world wars between 1914 and 1945 effectively brought the Polish and European amber industry to a complete standstill, with every craft workshop in Gdańsk having been destroyed during the fighting.

Immediately after the Second World War, the trade and manufacture of amber goods remained difficult, design being determined by a period of so-called socialist realism imposed by the communist Soviet government. The Factory of Amber Artefacts in Gdańsk controlled production and design of works in amber, using imported amber from the Soviet Union for many years. The death of Stalin in 1953, however, allowed a certain softening of the tight political control over artistic design. Although the Republic of Poland was born in 1952, it was not until the early 1960s, during more liberal times, that creative artists took an interest in amber again. Artists were once again able to demonstrate their skills in a prestigious museum with the opening of a permanent amber exhibition at Malbork Castle in 1965. At about the same time in 1966, museums such as the Muzeum Sztuki in Lodz, and the Galleria Foksal in Warsaw, supporting the avant-garde movement in Poland, were also opened.

Once again, Poland is a world leader in the production of unique works of art using their much treasured fossil resin. New techniques and new technologies, many of which are kept as trade secrets, are constantly being developed, compensating for those that may have been lost over the millennia of amber exploitation and political change in the region. In

The human-powered 14th century Great Crane in Gdańsk with shops selling amber on the left.

A piece by Giedymin Jabłoński.

1989, Poland rejected communism in favour of democracy with the election of the Solidarity Party. Artists were heavily influenced by the political changes that were taking place in the eighties and helped to develop the Polish consciousness of that period.

Poland has now grown to become one of the healthiest economies of the post-communism countries, and amber has contributed to this success.

Amber continues to be an important component of the arts, with festivals and exhibitions of modern pieces throughout the Baltic region. Artists from all over the world take part in competitions, exhibitions and workshops. There are now many contemporary amber art exhibitions worldwide displaying the work of students and established artists alike. One of the better-known amber festivals is Amberif (International Amber Fair) in Gdańsk, which is primarily for the amber trade. In March 2009, at the 16th annual Amberif, there were over 450 exhibitors from 12 countries. As part of this fair, there are two major competitions for designers: the thematic Electronos Amberif Design Award, and the

Teething necklace of raw Baltic amber from Lithuania.

Mercurius Gedanensis Jewellery Competition. There is also a Polish national competition for trainee and professional goldsmiths to promote the use of amber in jewellery, called the Amber Handicraft Award. In fact all the awards were created to promote the use of Baltic amber in jewellery and other artistic works.

Another major Baltic amber event is the four-day Amber Trip jewellery exhibition held in Vilnius, the capital city of Lithuania. Although it only began in 2003, it has already become well established and promotes Vilnius as the capital of the Baltic jewellery industry.

The most common use of amber in art is probably in jewellery, and one of the better-known jewellery artists from Gdańsk is Giedymin Jabłoński. He has been promoting Polish jewellery art and the use of amber in art since the early 1970s, at a time when few considered jewellery design an art. His work is varied and draws from a multitude of techniques and materials emphasising the strong link between jewellery and the wearer. Amber, being one of his favourite materials, is often used in his designs.

Jacek Ostrowski's Mercurius Gerdanensis winning design from Amberif 2006.

An ordinary person firstly associates the world of jewellery with shining precious pieces of jewellery and measures their value and 'beauty' by carats of precious metals and stones. Nevertheless, an artist, in particular the one who engages in jewellery for self-expression purposes, considers a piece of jewellery a work of art that may be worn (Giedymin Jabłoński).

As with all art, the work of the jewellery artist reflects their environment; changing fashions, trends, and beliefs all influencing their art. As communism in Poland collapsed, many people found it difficult to cope with the new freedoms. It was not so much that boundaries disappeared, but that boundaries shifted. As the artists found themselves with new boundaries, they produced works that were exploring and testing the limits of the new society and, in some cases, perhaps even defining the limits. The works of art that result from such an environment are sometimes ostentatious, rude, extravagant, elaborate, as well as simple, creative, and meaningful. With jewellery art, it is also essential that it is functional, personal, and that it communicates a message from the artist and possibly the wearer, although many pieces do hold a purely aesthetic pleasure. Jewellery is made to be worn and is therefore a powerful means to communicate ideas and feelings to a wider audience, and amber today is playing a major role in this form of artistic expression in the Baltic region.

The variety of forms of expression using amber as a medium of communication in art is great. There have been works of contemporary amber art that would appeal to all tastes and beliefs — some of these to only a small section of the community, and some that may not be considered as art at all by most people.

Not all amber is used to produce 'amber art'. In fact the bulk of amber used in jewellery is not considered to be art, but purely functional and of aesthetic value. Sculptures, carvings, brooches, rings, necklaces, earrings, and bracelets made, at least in part, of amber, are commonly sold as everyday jewellery. Mass-produced popular designs are sold worldwide. The amber may be shaped, treated, or even faceted almost as if made in a production line factory. Some pieces of mass-produced jewellery may even be 'one-off' designs if the raw amber or natural shape of the amber is unique. Beads are mass-produced from the raw, through polished, to shaped and faceted beads of different sizes. Cabochons of amber are often used in brooches, pendants and rings. The amber may be treated to produce the popular 'sun-spangles' effect, or welded together in a mosaic effect. Raw and polished amber necklaces are still being produced for teething toddlers, as the amber is softer than the teeth and is non-toxic. It is also believed by some to have health benefits for their children.

Amber extraction in the Baltic has decreased substantially in the last few years. In 2000, the estimated total weight of amber extracted in the Baltic region was about 510 tonnes, falling to an estimated 200 tonnes in 2006. The true figure can only be estimated because it is unknown exactly how much is extracted illegally. The estimated illegal extraction in 2000 was about 11% and had grown to about 30% by 2006. Over this same time period, Russia and Ukraine combined produced over 90% of the world's amber. Other non-Baltic amber-producing countries, such as the Dominican Republic, Mexico, Myanmar, Sicily and Japan, produced a total of about 2 tonnes of amber in 2006, a small fraction of that produced by the Baltic states. Increases in the environmental protection standards in mining operation in Russia and Ukraine have restricted the amount of amber available for jewellery production in Poland. The reaction of the Polish jewellery industry has been to increase the percentage weight of silver in amber jewellery from about 85% in 2000 to about 90% in 2007, so that the jewellery would hold its value. The increase in the cost of silver is also an important factor in the profitability of these amber products. In 2004, when Poland joined the European Union, silver cost $5.48 per ounce, which rose to $17.92 per ounce in 2008. This means that it is difficult for craftsmen and jewellers to maintain a stable price for amber jewellery without substantially reducing their profits, or making no profit at all.

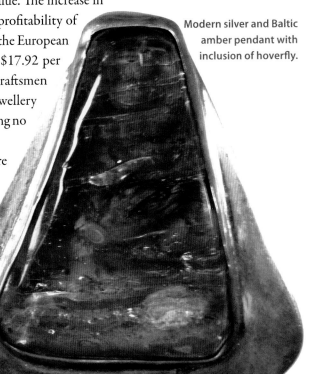

Modern silver and Baltic amber pendant with inclusion of hoverfly.

Amber production in Poland has become more industrialised in recent years. Larger companies benefited from a low flat rate income tax of 19% allowing them to invest in new facilities, equipment and shops. Smaller family businesses lost out, however, and the collapse of many of these small industries resulted in a huge reduction in those employed in the Polish amber industry.

Poland is still the major producer of finished amber products, with a total production of about $360 million in 2006. Lithuania is the closest competitor with only $100 million of finished product. Although these figures may be an underestimate, they provide an indication of the scale of the share of the finished amber products in the Baltic region. What they don't take into account are the large number of illegally imported finished products from Russia and Lithuania that are sold in the Polish market. The domestic amber product market in Poland is growing, with amber and silver products being most popular. It is not only the jewellery market that is healthy, but amber is also becoming more commonly used in luxury tableware, cutlery, and decorative objects. However, raw and small polished nuggets and cheap souvenirs are less in demand by the domestic market and are more likely to be traded abroad.

Although illegal trafficking of amber has been on the increase over the last decade, the Polish General Customs Inspectorate is taking it very seriously. Rather than targeting the small-scale smuggler, they are concentrating more on the larger quantities being smuggled across borders. In 1998 only a few dozen kilograms of illegally imported amber was seized at the Ukraine–Poland border. This rose to over 500 kilograms in 2006 and has since reduced, illustrating the effectiveness of the scheme. Rather than raw amber, the smugglers turned to worked amber, usually of high quality. This suggested that the smugglers had access to well-equipped illegal workshops and were able to hire skilled staff. Changes in the laws of the Ukraine may mean that these workshops are no longer illegal and may offer some competition to the Polish workshops, or even form partnerships with them.

Amber production is falling, and ageing mining equipment has not been replaced as the costs continue to rise. This may make it more and more difficult to obtain raw amber and could affect Poland, which imports a great deal of the raw product from neighbouring Baltic countries to supply its amber workshops and jewellery-making factories. Despite this, it is still possible to purchase amber jewellery on the internet quite cheaply. There are hundreds of internet amber shops, or amber shops with an internet presence. Buyers will always have to be aware that there are a lot of fakes available on the internet market.

Medicinal use of amber

Ever since man discovered amber in prehistoric times, it is likely that it was used in some way as a medicine. The Greek physician Hippocrates was one of the first to record the use of amber in medicine, but it is likely that it had been used in this way for a considerable time before that. The philosopher and poet Aristotle also recorded some of the medicinal attributes of amber in about 350 BC. It is not unknown for other fossils to be used as medicinal cures. In the early eighteenth century, the author Martin Martin recorded Jurassic age ammonites and belemnites being infused in water to relieve dysentery, diarrhoea, tuberculosis, worms and cramps on the Isle of Skye in Scotland. In fact, still on the Isle of Skye, a piece of amber, now in the National Museum of Scotland, was also used to relieve failing eyesight until at least the eighteenth century, by rubbing it on the eyelids. There are several beads and amulets of amber (called *lammar* in Scotland) that were used to cure a variety of ills. One from Argyllshire was recorded as being used to cure poor eyesight, and another one as a Highland charm to cure cattle from a host of diseases. Other countries, in Europe and elsewhere, also have a history of using amber for curing illnesses in a similar manner.

Amber has also been popularly used in the mouthpieces of pipes and other smoking aids. In the Far East it was believed that using amber would prevent the transmission of infections. It was also used in pipe mouthpieces because it does not have an odour or taste at room temperature, and emits an acquired pleasant smell when burnt. The amber excavated from mines was most highly prized by smokers for

Amber bead used to heal failing eyesight on the Isle of Skye.

© National Museums of Scotland.

Meerschaum pipe with amber
mouthpiece (19th century).

this use, as it was harder and clearer than the amber derived from the sea. Cheaper imitations using materials other than amber, such as Bakelite, Lucite, vulcanite, bone and horn were commonly used in the stem, but these were not quite of the same quality and not as highly prized as those pipes with the pure amber stems and mouthpieces. The only problem was that the amber, being brittle, broke easily, and hence many of these have not survived to the present day, and good examples are therefore quite rare.

AMBER CURES IN ANCIENT TIMES

From as early as the fourth century BC, amber was used for a variety of illnesses such as stomach ache, asthma, *delirium tremens*, whooping cough, and thyroid problems. According to Callistratus of Athens (sixth century BC), when amber was worn as an amulet by a child, or taken as a drink, it could be used to prevent delirium or cure kidney stones or a bladder infection. He also suggests that it be used as a cure for fevers and other diseases when worn as a necklace, and when powdered and mixed with honey and oil of roses is good for ear infections. Attic honey (honey of the sacred bee of Attica, Greece) was often used for improving eye sight and frequently mixed with amber powder to give it an extra boost. As a powder, sometimes with mastic gum in water, amber was also taken as a remedy for diseases of the stomach.

MIDDLE AGES AND LATER

In Medieval times amber worn as beads was used in the treatment of jaundice. The amber was thought to absorb the yellowness of the skin and reinvigorate the body. In the thirteenth century, the Dominican Bishop Albertus Magnus placed amber as the first of the six most effective medicines of his time. It was mixed with alcohol and used as a cure from stomach and rheumatic aches.

Lithuanian tribes were known to use incense to drive away evil spirits, and newborn babies were fumigated for similar reasons, as well as to make them grow faster. In the early Christian church, the use of amber in incense was adopted, presumably also to drive off evil spirits and imbue the congregation with a sense of wellbeing. The Christian church, Islam, and other religions also use amber in the form of rosaries or prayer beads.

Amber charm beads that belonged to the McDonalds of Glencoe.

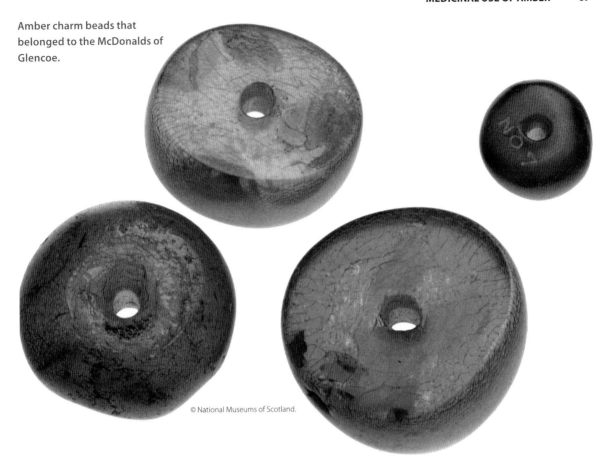

© National Museums of Scotland.

In 1502, Camillus Leonardus, an Italian astronomer, mineralogist and physician, suggested that if you place amber on your wife as she sleeps, 'all her evil deeds will be revealed'. It is not clear from his writings how this manifested itself, but it is hard to imagine that it would have worried too many wives!

At the same time as the amber business was beginning to flag in Prussia in sixteenth century, it was popular for the women of the east coast fishing villages of Scotland to wear a necklace of amber beads. They also hung amber beads on their children to protect them from evil. The MacDonalds of Glencoe had a set of amber beads that were used to cure blindness. Elsewhere amber beads were also used as a cure for sore eyes and sprained limbs. In about 1575, on the Buchan coast, a huge piece of amber was reputedly washed ashore: '...*arrivit ane gret lomp of this goum in Buchquhane, als mekle as ane hors*' (...arrived a large piece of amber in Buchan, as big as a horse). Although the text was originally written in Latin by the Bishop John Leslie of Ross in his book on the history of Scotland of 1578, it was translated into the Scots language by Father

James Dalrymple in 1596. It may be that the Latin text was mistranslated and that it just said that there was a truly big piece of amber washed up on the shores of Buchan, rather than one the size of a horse!

For those who are Latin scholars, the original Latin text by Leslie reads: *Ingens quædam succini massa, equi magnitudinem superans in littore Boquhanico nostro seculo fuit reperta.*

In the mid-sixteenth century, the popularity of amber was dropping as the Protestant Reformation took hold. The lack of demand for rosaries forced a desperate Duke Albert of Prussia to look for other means of supporting the amber trade. He asked court physicians to investigate the medicinal properties of white amber. The resulting monograph was the first scientific publication on amber, published in 1551 and entitled *Succini Historia*. It suggested a variety of medical prescriptions using amber, for ailments such as: toothache, stomach-ache, rheumatism, and heart arrhythmia. Physicians seemed to prefer the white variety of amber as being the most effective. The publication gave a respected credibility to the uses of amber in previous centuries.

In the Polish regions of Kurpie and Cashubia, people decorated their houses with amber to ward off evil. Apparently, fumigation with the smoke of burning amber prevented the spread of the plague during the Middle Ages. In 1680, according to the priest Matthaus Praetorius, it was stated that 'not a single amberman from Gdańsk, Klaipeda, Kaliningrad or Liepaja died of the disease'. If priests believed that amber had that kind of power, it is not difficult to understand why so many local people would have followed their example.

Amber was also thought to cure kidney stones, and was championed by none other than the theologian, Martin Luther, in the sixteenth century. A recipe for this medicine was published by seventeenth century Irish scientist and philosopher, Robert Boyle (made famous by his work on gases) in his *A Collection of Choice and Safe Remedies.*

Boyle writes:

> *The Medicine of a Famous Empyric for the Stone.*
> Take Amber (clear and yellow) Sea-horse Pizzle and Niter, of each a like quantity, (*Note will*, in case of Ulcerated Kidnies, put half the quantity of Amber) and an eigth part of the Nitre (of Natural Balsam.) Pulverise each apart, and make them up into Pills with Chios (or at least clean *Strasburg* Turpentine) take five, six or seven Pills (of above ten to one ounce) Morning and Evening.

At the same time, the medical writer and self-proclaimed *Professor of Physick*, William Salmon, wrote in his *Family Dictionary* that the cure for 'falling sickness' was to take 3 ml of amber powder in 140 ml of wine once a day for up to eight days. Either that or let the patient receive the fumes of the amber as it heats up on hot charcoals while the patient sits uncomfortably above them on a commode. It is not difficult today to see why this treatment is no longer in use.

Prayer beads from Turkey.

EIGHTEENTH TO NINETEENTH CENTURIES

In late eighteenth century Scotland, a smuggler of the name of Carnochan from Galloway had an oval amber bead that he wore around his neck. It was over 2 cm in diameter and 1.3 cm thick with a silver ring through the perforation. The story goes that Carnochan removed it from a *bing o'eththers* (mound of adders) which were busy making the amber bead. In Scotland, amber was thought to have been made by adders in a similar way to that of the so-called snake-stones described by Pliny in the first century AD. The story goes that the snakes would writhe together in midsummer and produce a 'bubble' in the form of a ring. They would then hiss and blow, throwing the ring up into the air with their tails, upon which the ring would harden like glass. It was thought that whoever possessed such a stone would prosper and be able to heal ills. This is strikingly similar to stories told by many cultures, including that of the Lapps, of 'powerful' snake-stones.

Carnochan wore the amber bead on a ribbon around his neck and used it for curing *backgaun weans* (sick children), *elfshot kye* (diseased cattle), and *sick beass* (sick animals), and as a talisman to avert the effects of the 'evil eye'. For the bead to be used, it was dipped three times in water which was then given to the sick child or animal to drink. Carnochan lost it whilst digging for worms in his garden one disastrous day and his luck left him. His smuggling booty and hiding place were found and he died in poverty. Many years later, one of his grandchildren found it in the garden. To see if it was still possessed by healing powers, it was used in an attempt to cure 'Jean Craig's cat'. Sadly, the cat died, and so it was thought of no more use and was eventually given to Dr Robert de Brus Trotter of Perth. What happened to it subsequently is unknown, but there are several such beads in museum collections in Scotland.

In 1832, Anthony Todd Thomson described the uses of the oil of amber obtained by the destructive distillation of amber (heating amber powder to release and collect the residues). The oil, he states, has antispasmodic powers and was to be taken orally during periods of hysteria and other convulsive diseases. By the time he wrote this, however, this practice of taking the amber oil orally had gone out of favour. He does state that 'it is still, however, externally employed' for its antispasmodic powers. Amber oil was also used as the active ingredient in a concoction to cure whooping cough, but Thomson does note that it is not a very effective treatment when compared to belladonna and conium.

AMBER IN THE TWENTIETH CENTURY

In the early 1900s it was popular to consider amber as a panacea for asthma, oedemas, toothache and various other ailments. The oil of amber (*oleum succini*) was still being used in small doses for 'hysterical affections' (such as problems of the nervous system, epilepsy, and spasms). It was also used to warm the skin surface and for the relief of bronchial problems and rheumatism. Amber oil was still an ingredient in whooping cough medicines such as 'Roche's Embrocation' from the early 1800s until well into the twentieth century and was rubbed on the chest to reduce coughing spasms. With the development of a vaccine for whooping cough in the 1920s, and its widespread use in the 1950s, the ointment was effectively made redundant.

AMBER IN MEDICINES OF THE TWENTY-FIRST CENTURY

Rather than developed chemical-based drugs, alternative remedies claim to rely on natural extracts from plants, minerals and animals. Although their effectiveness is disputed by many conventional doctors, alternative medicines have become more popular in recent times.

Today homeopathic pharmacies in Poland sell amber ointment and amber spirit as a cure for rheumatic or neuralgic pains or arthritis. Another amber-based product is used for the temporary relief of hay fever and pollen-induced allergic symptoms such as sneezing, runny nose, itchy and watery eyes, sinus congestion and headache. Products such as succinic acid (obtained from amber) mixed with chitosan (hydrolysed shrimp cuticle) is also claimed to help with weight loss. Amber extracts have also been used to treat phobias such as siderodromophobia (the fear of trains and train travel) and claustrophobia (fear of enclosed spaces).

Succinic acid is in fact a naturally found substance in the body. In 1865, the Nobel Prize winner Robert Koch (a German physician) showed that with certain diets the body excreted large amounts of succinic acid. In one such experiment he ate half a pound of butter a day to show how fat affected the production of succinic acid in his urine. He became so sick after five days that he then decided to limit his study to other animals!

Recent research suggests that succinic acid helps improve the immune system, and a number of new medicines of this type have been produced. Some consider it to be an antioxidant which may reduce ageing of human cells. The Russians use succinic acid as a treatment of alcoholism, where they suggest that it may help in the recovery of damaged brain function. They also have incorporated this amber derivative into agriculture, where it is claimed that crop yields appear to increase and plants become more resistant to fungal and bacterial diseases.

Amber is taken in many forms: orally, it can be taken as a powder or a tincture; it can be taken as a suppository with honey; inhaled as smoke from burning amber; and also can be used on the skin as an ointment, oil, poultice, powder, or beads. The latter may be worn in the form of jewellery.

Interestingly, in China, amber is used for very similar purposes to that of its historical and homeopathic uses in Europe. It is used for its calming effect in the treatment of palpitations, amnesia, insomnia, and epilepsy amongst others. The tiger's soul (*Hupo*), or amber, is usually administered as a powder with other ingredients, depending on the illness. It is thought to relax the mind, and relieve convulsions. Amber is also used to help alleviate urinary problems such as stranguria, kidney and bladder stones, as well as lower abdominal pains and poor blood circulation, including heart disease. When mixed with pearl powder, it has been used as a tonic, to help with stomach pain, as well as an added aid to recovery for cancer patients. In Chinese medicine amber is also used for the promotion of tissue regeneration and in the treatment of ulcers, boils and other skin complaints.

An additive to medicines for almost all aches and pains, regenerations of injured tissues, skin complaints, allergies and much more, amber appears to have been used almost as a panacea of medicine. How effective it is, has not been fully investigated. Perhaps the amber guilds of today will fund studies into the health benefits of amber just as Duke Albert of Prussia did in the 1500s?

Amber in science

Since the time of Thales in ancient Greece (sixth century BC), amber has been a popular substance of intrigue for the scientist. What is amber? How did once living organisms get into the amber? What can these inclusions in amber tell us about life millions of years ago, and what does this tell us about their development over time? These are just a few of the questions that have been asked over the centuries.

AMBER STUDIES

Given that there are about 1 in 100 nuggets of Baltic amber containing an inclusion, the total number of inclusions that should be available each year for study would be about 2 million! This provides scientists with a huge amount of material on which to base their research and has resulted in well over 500 new species of fossil insects being discovered.

The first comprehensive scientific publications on the fossil plants and animals in Baltic amber were published in the 1840s and 1850s. Since then thousands of articles have been published on Baltic amber and over a thousand articles on Dominican amber since 1960. One of the earliest representations of the amber tree was published by Jacob Maydenbach in *Honorius Sanitatus* (1491) and even amber fakes were illustrated in 1583 by Daniel Hermann. In 1742, Nathanael Sendel in his *Historia Succinorum* illustrated over 250 pieces of amber with plant and animal inclusions. Even Robert Hooke, the 'Father of Microscopy', had an interest in amber and its origins. He reported on a piece he had seen with a fossil gnat and the moss of a tree in a private collection of amber. He also mentioned that Wigand von Marburg reported a green tree frog included in Baltic amber. No Baltic amber in any present day collection is known to have frog inclusions, and therefore this claim has never been substantiated.

One interesting aspect of the study of inclusions in amber is that they give us a clue to the diversity of life at the time the amber formed. Not all animals in the fauna will be preserved in amber, though. A study by David Penney, in 2005, comparing the fauna living on the island of Hispaniola to the fauna preserved in Dominican amber, suggests that amber contains only a small fraction of the total biodiversity of the amber forest. Not all the insects present in the ancient amber forest were as likely as others to get trapped in resin, as some were more closely associated with the resin-producing trees. Larger insects and animals were also less likely

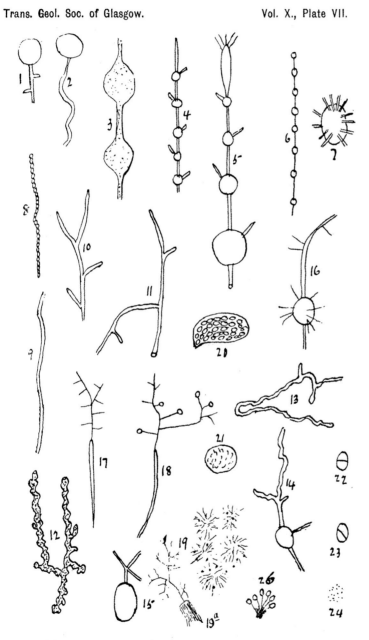

Trans. Geol. Soc. of Glasgow.   Vol. X., Plate VII.

MICROSCOPIC PLANTS IN AMBER (Middletonite) FROM COAL.

Inclusions in Carboniferous 'amber' from Scotland (Smith, 1894).

to be trapped in amber, as they may have found it easier to escape the sticky resin. By comparing the living forests of Hispaniola today with ancient life known from Dominican amber, scientists can make an educated guess as to what might be missing from the fossils to help them to build a more complete impression of life in the amber forests.

Air trapped in amber has frequently been analysed to provide a guide to the oxygen levels in the atmosphere at the time the amber was produced. Some studies suggested a higher level of oxygen in the atmosphere than the present day; others that it was the same. The chemical make-up of amber is a complex system where diffusion, solubility, and chemical reactions take place. Oxidation of amber is a clearly visible reaction, as it turns the golden yellow colour of amber to a darker red amber. A study by Thure Cerling in 1989 concluded that the oxygen content of amber does not reflect the composition of the palaeo-atmosphere, but rather reflects the ongoing interaction with the surrounding environment over time.

The oldest 'amber' (also known as middletonite) to have been found with inclusions was found from the coalfields of Ayrshire, Scotland. John Smith published this discovery in 1894, describing the inclusions as parts of coniferous plants and fungi. Although the actual specimens John Smith studied are now lost to science, new research using some of the more modern techniques like 3D X-ray imaging, which looks at opaque amber, may reveal evidence of such inclusions in this type of resin.

CHEMISTRY OF AMBER

Many attempts have been made to differentiate amber from different regions using chemical analysis in order to identify the ancient trade routes of amber. There are many factors that can influence the

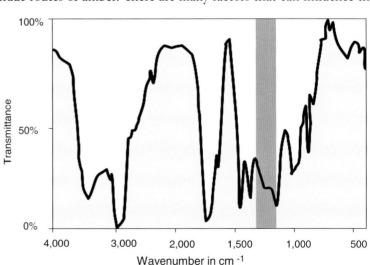

Graph from infra-red spectroscopy showing the 'Baltic shoulder' in orange.

chemistry of amber, including the type of tree that originally produced the resin. Temperature, pressure, exposure to oxygen, water levels, type of sediment, and the age of the resin can all also have an influence on its chemical characteristics. Amber from the same age and same place but from different sediments may be classified as different types due to the effect that different environments have on the amber. Resin contains alcohols, oils, and acids that go through a process of polymerisation to become copal and possibly eventually amber. Polymerisation is the process whereby bonds form between organic molecules within the resin to produce a more stable and permanent compound such as amber. The process may take several millions of years before the resin is stable enough to be called amber.

Amber has a fairly complex chemistry that is often very variable, which is due to the fact that amber is not a mineral, but a mixture of organic compounds derived from living matter. Even amber samples from the same locality may have slightly differing compositions. Amber is in fact a natural plastic and does not have a consistent chemical formula due to its variability. Carbon, oxygen and hydrogen are the main elements that make up amber, but sulphur is also present in small quantities, along with some other minor elements.

Baltic amber is unusual in that many samples contain up to 8% succinic acid. The succinic acid was perhaps produced as a by-product of the fermentation process of the tree resin during the Eocene in the Baltic region. It is also possible that it is produced naturally by the tree, as it occurs in many living plants and animals (although to a much lesser extent than Baltic amber). Baltic amber is not the only amber to contain succinic acid as it occurs, although to a lesser extent, in amber from Mayanmar and Sicily. Retinite is the name given to the forms of amber that lack succinic acid., whereas Baltic amber is also known as succinite.

Analytical methods such as pyrolysis, infra-red spectroscopy and thermal analysis are useful tools in the determination of the place of origin of a piece of amber, but are not unequivocal. Pyrolysis involves heating a piece of amber and analysing the vapours produced using a mass spectrometer. Different ambers produce different amounts of the organic compounds that are emitted on heating, making it often possible to identify their origins.

Infra-red spectroscopy is performed on powdered samples in a solution of potassium bromide (or other salt). Spectroscopy measures the absorption of various wavelengths of light in the infrared part of the spectrum. Most Baltic amber has a particular infrared spectroscopic graph profile that is characterised by a 'Baltic shoulder' (amber strip in diagram).

Another way of identifying different ambers is by thermal analysis. The change in weight of amber undergoing combustion is precisely measured in relation to changes in temperature from 200°C until combustion is complete (at about 600°C). The resultant weight loss curve can be used to identify ambers. Studies on ancient amber from the Upper Triassic (about 225 mya) to the Eocene Period (about 40 mya) showed that it was possible to distinguish between amber on the basis of age and geographical origin. It appears that ambers of different ages, having undergone different physical and chemical processes over time (diagenesis), and possibly from different palaeobotanical origins, have different weight loss curves.

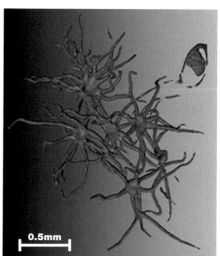
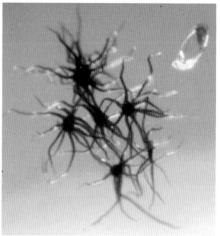

Confocal laser scanning microscope (CLSM) images of a Baltic amber trichome and a group of Mexican trichomes using CSLM (top right) and transmitted light (bottom right) – note air-bubble in top right corner.

0.5mm

0.25mm

NEW TECHNIQUES

Sadly, all the above methods of analysing amber are rather destructive. The only non-destructive method of identifying the geographic origins of amber is to identify the extinct animal and plant inclusions contained within. Observations of the chemistry, diagenesis and preservation of the inclusions are also very useful in distinguishing between the different ambers. Some Mexican amber, for example, appears to have been heated slightly, producing a ghosting effect on the inclusions. Baltic amber does not seem to have been affected in this way, but frequently has a white fungal growth on insect inclusions. Both these ambers contain microscopic plant hairs called trichomes, which look slightly different. The Mexican trichomes seem to be more flattened star-shapes and the Baltic amber has more 'hairy' trichomes, reflecting the types of vegetation that the trichomes came from. In the case of the Mexican amber,

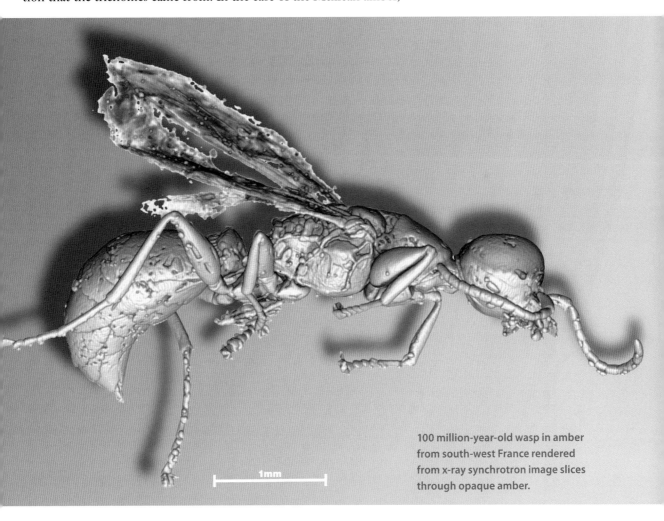

1mm

100 million-year-old wasp in amber from south-west France rendered from x-ray synchrotron image slices through opaque amber.

the trichomes have the appearance of those from some living species of the genus *Croton*. The Baltic amber trichomes have traditionally been ascribed to oak trees, due to the close association of these with fossil oak flowers also preserved in amber.

Future studies on trichomes trapped in amber should help confirm the identity of the source species for these. Confocal laser scanning microscopy has been used to examine these minute trichome inclusions in the deep infrared spectrum, producing high resolution three-dimensional images. Unfortunately, it is more difficult to examine insects in Baltic and Mexican amber using this technique, as both the insects and the amber fluoresce at similar wavelengths in the ultraviolet.

To view insects in amber, a light microscope is usually sufficiently effective, but is completely useless when it comes to opaque amber. To get around this problem, a group of palaeontologists in France, led by Malvina Lak and Paul Tafforeau and their colleagues, examined some opaque Cretaceous age amber from France using an X-ray synchrotron at the European Synchrotron Radiation Facility in Grenoble, France. The results of this study in 2008 were amazing. Not only were they able to find hundreds of insect inclusions in these ancient samples, but they were able to render computer animations of the inclusions to view them from any angle in high resolution. The overall resolution was better than anything possible using a normal light microscope. Large three-dimensional plastic printouts were made as a physical record of species new to science that would never have otherwise been seen in the opaque amber. The three-dimensional printouts are an accurate model of the insect trapped in amber, only over 300 times enlarged.

As new techniques for analysing and investigating amber are developed, new species of plants, insects, and other animals will continue to be discovered, and in greater detail — all building a more complete picture of what the amber forests may once have looked like, whether in the Baltic region, Dominican Republic, Myanmar, or elsewhere. The continued use of analytical techniques may also help determine the origins of amber, the tree that produced the resin, the changes it has undergone over time, or a more precise age for the amber. It is an exciting time for amber researchers and there is no telling what discoveries the next decade will reveal.

Amber and the dinosaurs

A question that is frequently asked is whether you can bring dinosaurs back to life from amber?

In Michael Crichton's book, and the subsequent Steven Spielberg film *Jurassic Park* released in 1993, it was suggested that dinosaurs could be recreated from DNA taken from mosquitoes in amber. The idea was that a mosquito, having just fed on the blood of a dinosaur, then got caught in the sticky tree resin, preserving both the insect and its blood meal containing the dinosaur DNA.

There are several problems with this idea. The first is that there are no known insects in Jurassic amber. Secondly, there have been very few mosquitoes ever found in amber of any age, and the chances of it just having fed on a dinosaur are fairly remote. There are, however, some biting insects that are from the age of the dinosaurs. Blood-feeding black flies are known from Jurassic rocks as well as in Cretaceous amber, and horse flies have been found in Cretaceous amber from Dorset in England. Biting midges are also known, and some even had mouth parts that could have bitten through the thick skin of a dinosaur. Hence there are some candidates other than mosquitoes which could potentially provide scientists with dinosaur DNA.

Theoretically then, we might not be able to create a *Jurassic Park*, but could surely recreate Cretaceous dinosaurs such as T. rex? Unfortunately it is not that simple. Assuming that an insect bit a dinosaur immediately before it was trapped in resin, the gut flora and stomach acids of the insect would continue to digest the blood after the insect itself had died. This means that the precious DNA of the dinosaur would be at least partly digested and broken down. In truth, most insect specimens found in amber, particularly Baltic amber, are in fact hollow shells, as the internal soft tissues have rotted away before the resin turned to amber. It is such a huge challenge to extract DNA from the insect itself in amber; so imagine how much more challenging it will be to extract DNA from the digested gut contents of its stomach.

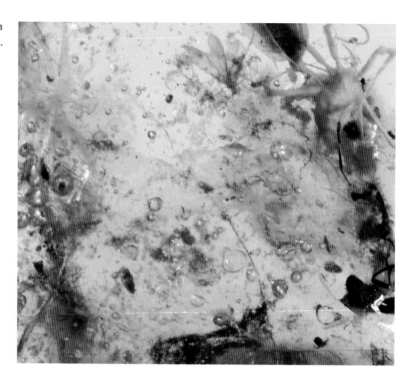

Even if scientists did find an insect specimen that was so well preserved that it still had its soft tissues intact, what are the chances of them
recovering DNA from it? In an attempt to answer this question, several
scientists from the Natural History Museum in London took some
insect specimens trapped in East African tree resin called copal, which
was possibly as young as 50 years old, to see how the DNA survived
over the shorter timespan. They also looked at some insects in Dominican amber that were over 40 million years old from which DNA had
previously been reported. Neither the insects in the Dominican amber,
nor the insects in the East African copal, produced any insect DNA.
They hence concluded that it was not worth destroying valuable fossil
insects, which are important to evolutionary science, in order to attempt
to extract DNA. Of course the absence of DNA in these specimens does
not mean that it does not exist, and it is likely that some scientists will
continue searching for DNA from amber.

Let's assume that at some point in the future scientists do manage,
using new techniques, to extract DNA from the gut of a biting insect in
Cretaceous amber. How easy would it then be to build a dinosaur?

Many people will be familiar with the famous use of DNA molecules in the cloning of Dolly the sheep in 1996. So surely if scientists
have the knowledge and technology to clone one animal they can clone
any animal? However, for cloning to take place scientists need to have

a complete animal cell containing the intact DNA of that animal. The story of the Woolly Mammoth illustrates just how difficult this is to achieve. In 2007, a 37,000 year old baby Woolly Mammoth was recovered from a frozen sandbar on the Yuribey River in north western Siberia. Russian Scientists had rejected the idea of trying to clone the mammoth, as complete cells are required for cloning and the mammoth's cells had burst under the freezing conditions of the tundra ice. However, over 70% of the DNA from the hairs of several mammoth specimens has been successfully sequenced in 2008 by American and Russian scientists.

Contamination is another major problem in sequencing this type of ancient DNA. It is possible that some of the supposed mammoth sequence is actually from bacteria and fungi involved in the decay of the mammoth corpse. Using hair makes it easier to remove a lot of these contaminants, but not all of them. Comparing the proposed mammoth DNA to that of an African elephant, its closest living relative, may help scientists to identify some of the contaminant sequences. This project is still ongoing.

Despite these initial problems, Japanese scientists have been attempting to clone a Woolly Mammoth for a number of years, but have not yet been successful. This is because the DNA is too fragmentary, and possibly contaminated, making it difficult to match sequenced sections. Recent research, however, may make cloning the Woolly Mammoth a possibility. In 2008, Japanese scientists based in Kobe were able to clone mice from brain cells that had been frozen for over 16 years. It is thought that high sugar levels in the brain may have protected the DNA in the cells from destruction. Perhaps the use of cells from a frozen mammoth's brain will hold a key to cloning mammoths.

There are an incalculable number of hurdles to cross before scientists can even begin to consider recreating an extinct animal and, even more so, from DNA in amber. It may be that the mammoth genome is nearing completion,

A 40,000-year-old mummified baby mammoth from Siberia nicknamed Lyuba.

but we can only imagine what further scientific complications may present themselves as attempts are made to clone from these nearly complete sequences. Perhaps sometime in the near future, we will see the first clone of a Woolly Mammoth. This might raise an ethical debate as to whether it is right to clone an extinct Ice Age mammal, and at a time of global warming.

Although, potentially, mammoth clones are possible, dinosaurs are another story entirely. Firstly, there are no existing ice sheets that would date back to the time of the dinosaurs, and thus scientists will never find a source of DNA from a frozen carcass. Secondly, if DNA trapped in amber was from a dinosaur, how could researchers identify it as such, and what living animal would be the best model for comparison? In order to identify dinosaur DNA fragments, scientists would need to look at birds, such as the ratites (ostriches and emus), which are their closest living relatives. But the bird DNA would contain significant differences from dinosaur DNA, considering the 65 million year gap in evolution of their genome.

Contamination is still a serious issue that faces anyone who claims to have recovered DNA from any fossil material, including amber. It will be enormously difficult to have any degree of certainty as to exactly what the sequences represent. In fact, if we cannot obtain viable DNA from solid bone and tissue that is less than a hundred thousand years old, what chances do we have in obtaining good DNA from over 65 million year old amber? The answer from most scientists is likely to be 'virtually no chance at all'.

So it looks unlikely that dinosaurs will be recreated from the guts of fossilised blood-sucking insects in amber. In fact, to answer the original question 'Can you bring dinosaurs back to life from amber?', I think it would be safe to say: 'No ... at least not in our lifetime'.

The amber trap

One of the most exciting aspects of amber is that it contains the fossil-ised remains of plants and animals in a beautiful state of preservation. These inclusions were even noticed by the Romans who brought some large pieces of amber to Rome for Nero's gladiatorial games. They were recorded by Pliny the Elder in the first century AD:

> The heaviest piece brought to Rome by Juliannus weighed about 13 pounds. That amber first exuded as a liquid is proved by certain objects, such as gnats and lizards, being visible inside it. These undoubtedly stuck to the fresh sap and remained trapped inside when it hardened.

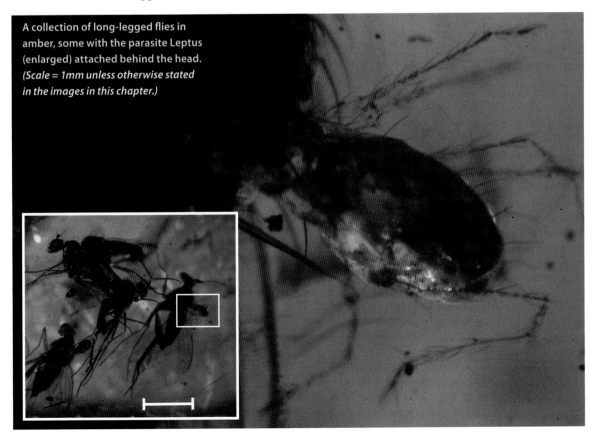

A collection of long-legged flies in amber, some with the parasite Leptus (enlarged) attached behind the head. *(Scale = 1mm unless otherwise stated in the images in this chapter.)*

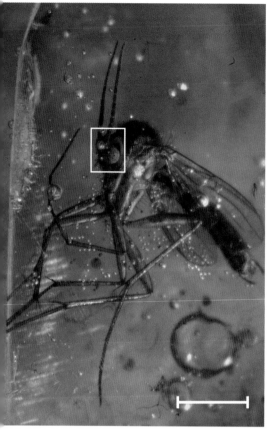

Fungus gnat in Baltic amber showing enlargement of the head.

Although it is inevitable that amber inclusions were known of in pre-history, there is no evidence to suggest they have been valued over pieces without inclusions. However, in 1868, an article written for Victorian Britain suggested that amber containing inclusions cost substantially more than those pieces without:

> Some amber is so discoloured by substances that have adhered to and become mixed with it before being hardened, that it will not bring in the market more than four or five shillings per pound. Other pieces that are clear, or that can be used as specimens, containing insects in perfect shape, are worth from £16 to £17 per pound.

This is equivalent in today's money to £35 per kilo for the low grade amber, and over £2,250 per kilo for amber with inclusions. On the internet in 2009 it was possible to bulk purchase small pieces of raw amber for about £300 per kilo. Amber with known inclusions was more difficult to assess, but depending on the inclusions it could be valued at between £2,000 per kilo for the more common fossil inclusions, up to over £60,000 for rare individual amber pieces with unusual inclusions, such as lizards, feathers, or mating ants.

Whatever the monetary value of the amber with inclusions, the value to Science and our understanding of the ecology of the 40 million year old Baltic amber forest is incalculable. The inclusions tell us a great deal about the animals and plants that lived in the amber forest, changes in climate and the seasons, interactions between the flora and fauna, and important clues to the evolution of certain plants and animals.

TYPES OF INCLUSIONS

The term 'inclusion' refers to just about anything that is encapsulated by the amber. There are many different types of inclusions recognised in amber. These range from vertebrates (such as frogs and lizards), insects and other invertebrates, fungi, bacteria, and plants, to bubbles containing water and air, faeces, minerals, and other drips of amber. For the most part, inclusions in amber refer to the remains of plants and animals enclosed within.

Plants

As amber is the resin of a forest tree, it might seem reasonable to assume that the vast majority of inclusions are of plant material, and more specifically of the plant that produced the amber resin. Surprisingly, however, this is not the case. There are a wide variety of plant remains found in amber. One of the most common of the plant inclusions are the stellate hairs, or *trichomes*. These minute star-shaped inclusions are very common and are a characteristic of Baltic amber. Stellate hairs can offer defences against insects and larger herbivores as well as give protection from wind and rain, which is a potential sources of damage to the softer tissues of the plant. They can be found on the leaves, stem and flower buds of many plants today.

The leaves of plants are also relatively common inclusions, but are usually fragmented, or partially decayed. In Baltic amber, at least 26 different types of plants have been recorded. When put together, the types of plants that are found in Baltic amber suggest that the forest is typical of a tropical to subtropical climate, similar to that found in North Africa and southern Europe today. The most common types of plants represented in Baltic amber are oak trees and the conifers, such as cypress and pine. Other, perhaps less common, types include willow, date palm, sorrel, hydrangea, geranium, juniper, magnolia and camphor.

A variety of plants found in Baltic amber. a) a twist of birch bark?; b) a pine needle; c) a fragment of moss; d) a cypress flower.

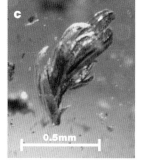

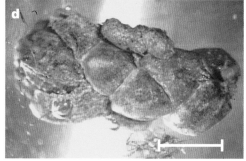

Stamen of a flower in Mexican amber.

Occasionally, the flowers of plants are also preserved in the amber, the flower of the oak tree being one of the most common types of this inclusion. Flowers of other plants do turn up, though, and representatives of the rose, witch hazel, pepperbush, and tea plant are known from Baltic amber. Rarely, fruits of the woodsorrel are also found. Other plant inclusions include the liverworts and mosses, and ferns.

Although no mushrooms have yet been described from Baltic amber, there are examples known from the younger Miocene amber from the Dominican Republic and the older Cretaceous Period amber from New Jersey.

Animals

By far the most common inclusions in Baltic amber, after the stellate hairs, are the insects and spiders. Based on the Baltic amber collections of the Hunterian Museum in Glasgow, a rough estimate of the proportions of different types of animal inclusions suggests that about 43% of animals are dipteran flies and over 20% are spiders. Bees, wasps and ants only account for about 9% of the total, springtails also make up about 9% of the fauna, and beetles only 6%. The remaining 13% is taken up by the less common representatives of the fauna such as cockroaches, crickets, and worms.

Some of the rarest of the animal remains in amber are the vertebrates. Although whole mammals have never been found, mammalian hair and some bones have occasionally turned up. One interesting study on Dominican amber has even been able to identify which group of mammals some hair belonged to. Poinar, a scientist from the University of California, identified two parasites associated with this clump of

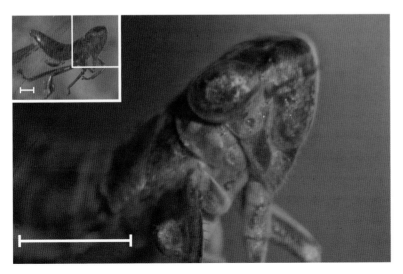

Some animals are caught at different stages of their growth cycle. This is a nymph of a leafhopper with enlargement of the head and mouth parts.

Tuft of fine mammal hair in Baltic amber.

The "Gierłowska's Lizard" (*Succinilacerta succinea*) – one of the rare lizard inclusions found in Baltic amber. This specimen is from the Amber Museum's collection in Gdańsk (a branch of Gdańsk History Museum). The item was purchased and donated to the museum by the Leopold Kronenberg Foundation of the Citi Bank Handlowy in Warsaw.

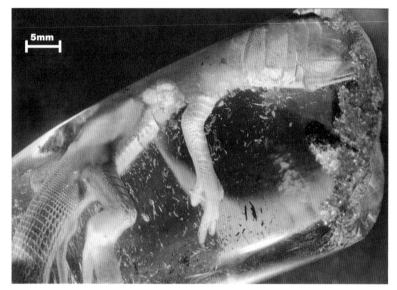

around 50 hairs. The physical characteristics of the hair had suggested that they might be from a rodent, and when identified, the associated mite and a beetle larva were found to be the same as contemporary parasites of modern-day rodents. The bones of a shrew-like insectivore have also been found in Dominican amber. How these hairs and bones became trapped in amber is unknown. It may be that they were dropped by birds, or in the case of the hair, the animal may have rested against the amber-producing tree, leaving some hair in the sticky mass of resin as it pulled away.

Bird feathers are also occasionally found from Baltic amber, but probably the best known of the rare fossils are the lizards. The first recorded lizards in amber were recorded by the Romans in the first century AD. However, it was not until as late as 1889 that a lacertid lizard, or true lizard, was scientifically described. This lizard, which came from Kaliningrad, went missing along with other museum collections during the Second World War, but eventually turned up safe with the other collections at Göttingen University in Germany. During the period it was missing there was much discussion as to its authenticity as a piece of Baltic amber, as a similar lizard was described in much younger African copal. It was finally validated as being Baltic amber in 1998 when the lizard was afforded the genus name *Succinilacerta*. Additional material from three different species of *Succinilacerta* have been described, including the Gdańsk lizard discovered by Gabriela Gierłowska, and a species of gecko from Kaliningrad in 2005. These are very rare occurrences, and because of their rarity and consequent value, lizard inclusions are commonly faked.

IDENTIFYING INCLUSIONS

To be able to properly identify the huge diversity of amber inclusions it is often nessesary to employ the skills of a specialist scientist, or a very experienced amateur collector. There are hundreds of scientific articles and books which cover the topic of amber inclusions, and it can take many years to become expert in this area of research. The next few pages will provide an insight into some types of animals and other inclusions that can be encountered in Baltic amber. Some of these inclusions are relatively common and some are very rare, but unless you can accurately identify the inclusion, it is impossible to tell which. There are books that help the amber collector identify inclusions more precisely, and some that provide identification keys for both amateurs and professionals alike. Here, the inclusions are presented as a series of images showing details of what can be seen through a microscope or magnifying glass. Sometimes it is very difficult to identify inclusions because the amber has cracked, or is full of bubbles covering crucial features, or is obscured in other ways. Inclusions such as insects can be lost or partially damaged by the polishing process, and by natural erosion on the seabed. Often there is only a fly's leg, or a wing that has been torn from the body as the animal has attempted to escape its trap. Partial specimens are, in fact, much more common than complete animals.

Normally, if amber inclusions are being photographed for publication, the amber is cut flat close to the inclusion, and at the desired angle

Midge in Baltic amber.

to obtain the best image. This technique can be very damaging to associated inclusions. A scientist with an interest in insects may decide to cut through the amber in a way that damages plant life or fungi, which may be of scientific importance to other fields of research. In the following series of photographs, the amber was not cut in this way and was only given a good polish. The amber nuggets were a variety of shapes and sizes and the inclusions were at various depths and orientations to the surface. In order to be able to photograph these inclusions and eliminate as much internal reflection and distortion as possible, the amber nuggets were immersed in clear oil that had a similar refractive index to amber (about 1.54). Although a mineral oil, and therefore not suitable for the long-term conservation needs of amber, Johnson's Baby Oil has a similar refractive index to amber and is quite good at eliminating the distortions caused by refraction and surface reflections.

COMMON AND LESS COMMON INCLUSIONS

Identifying inclusions in amber is not always easy and it is often the case that only parts of the animals or plants are preserved. Fractures, fungal growth and bubbles can also make it difficult to be certain what the inclusions are. On the rare occasion when the inclusions are well presented, it may still not be easy to identify them without expert knowledge and a comprehensive reference library of illustrations and descriptions. The photographic images on the pages that follow are a selection of inclusions from the collections of the Hunterian Museum in Glasgow. The amber pieces are not cut and polished for scientific study but are photographed through various thicknesses and curves of the amber pieces. They provide a guide to the types of preservation, the natural poses of the inclusions and a basic identification of some of the more common types.

Opposite: A selection of the more commonly encountered animals found in Baltic amber. From top left: a) cicada; b) dwarf weaver spider; c) cobweb spider; d) aphid; e) springtail; f) dwarf six-eyed spider; g) mite; h) silverfish; i) fungus gnat; j) caddisfly; k) long-legged fly; l) weevil; m) marsh beetle; n) wasp; o) ant; p) bee.

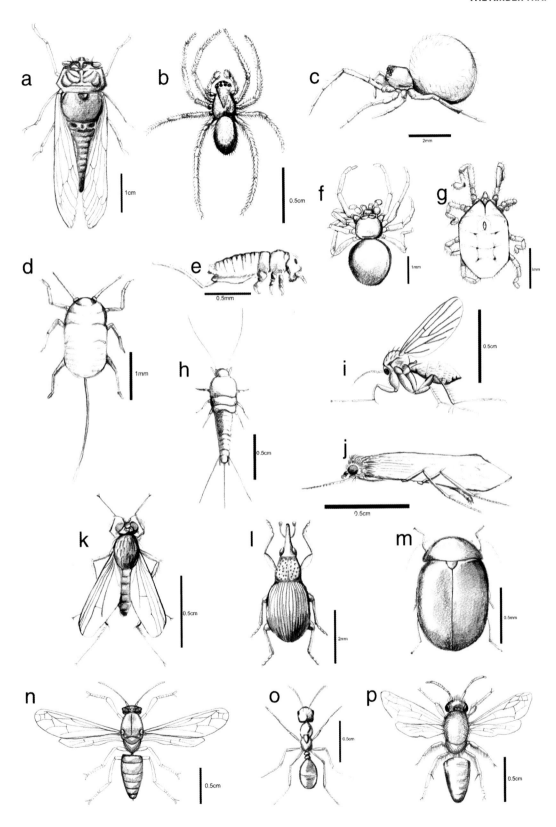

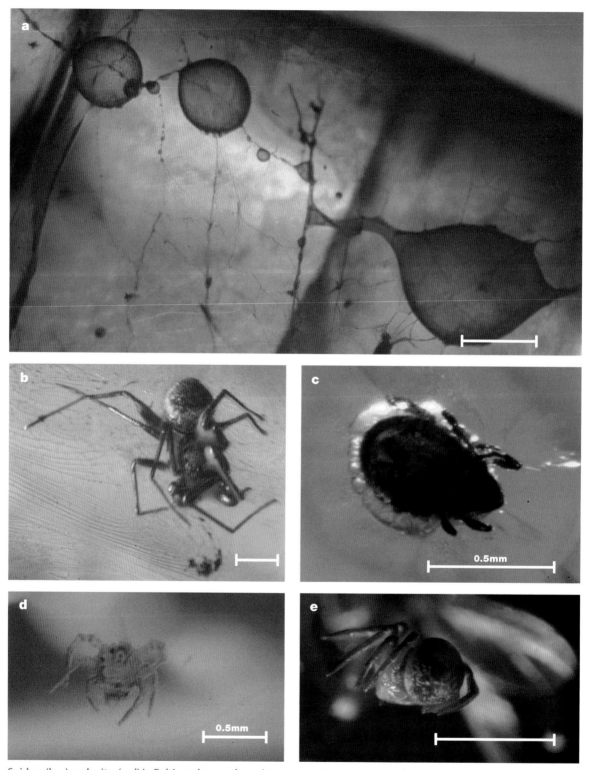

Spiders (b, e) and mites(c, d) in Baltic amber can be quite common in some pieces of amber, but a spider's web (a) is very rare. The web material is strong enough to survive the amber flowing over it, sometimes preserving trapped insects or even dew.

Two of the larger animals found in Baltic amber, a cockroach (a) and a harvestman spider (b), both of which show the milky product of fungal decay of the animals. This is a common characteristic of larger animals in Baltic amber, including some flies and centipedes.

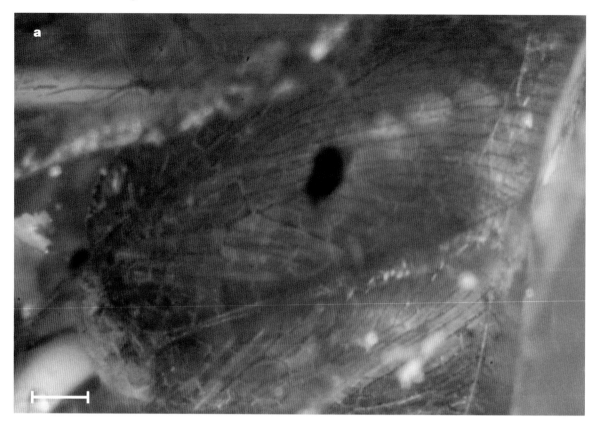

Not all large animal inclusions have the fungal growth obscuring the fine details. This lacewing (a) and cockroach (b) are particularly well preserved, although the cicada is missing the back end of its wings at the edge of the amber nugget.

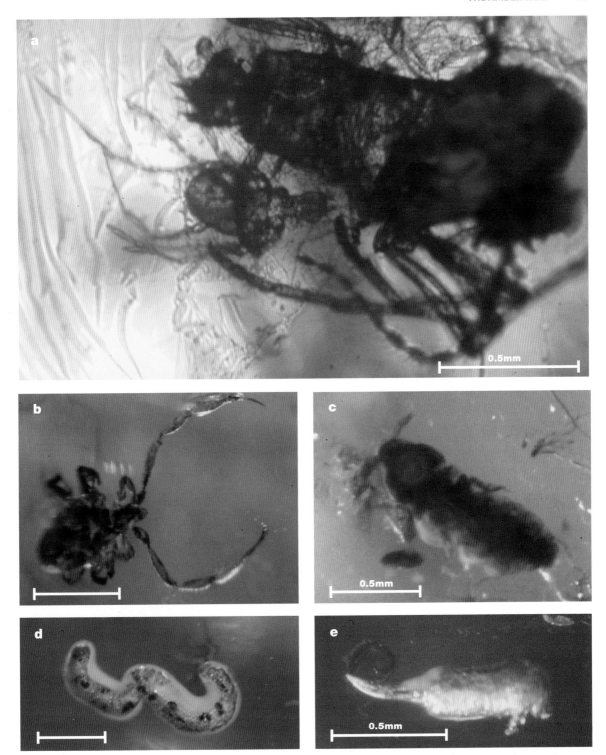

This selection of unusual animals shows a parasitic mite (Leptus) attached to a small mothfly (a). Around 25 species of pseudoscorpions (b) have been found in Baltic amber, but are still fairly uncommon. Despite being quite common, springtails (c, e) are very small inclusions and rarely noticed by the casual observer. This segmented worm (d) is extremely rare and is yet to be identified.

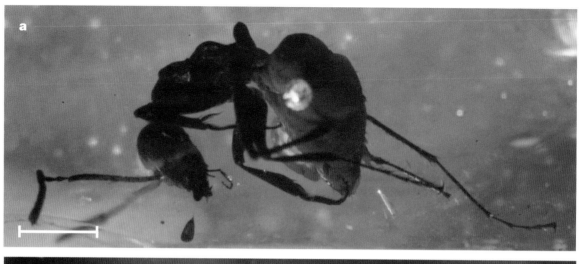

Some insects are seen to be struggling to pull themselves free from the amber trap, like this contorted ant (a) and aphid (c), whereas others, like the click-beetle (Pactopus) (b), draw their limbs in until the danger has passed, unaware that the amber had totally encapsulated it, and that it would be trapped there for the next 40 million years.

Multiple inclusions are not uncommon; for example the rove beetle above two long-legged flies (a). Cracks sometimes appear through fossil insects, perhaps because the amber is weakened where decay has taken place; such as is the case with this spiny weevil (b), or at the edge of a flow of amber resin over another, such as with these parasitic wasps (d, e). The underside of this marsh beetle (c) seen here is unusually well preserved for a beetle.

This enlargement of the head of a long-legged fly in Baltic amber shows just how much detail is preserved of these insects in amber. Each lens in the eye and sensory hair on its head can be clearly seen.

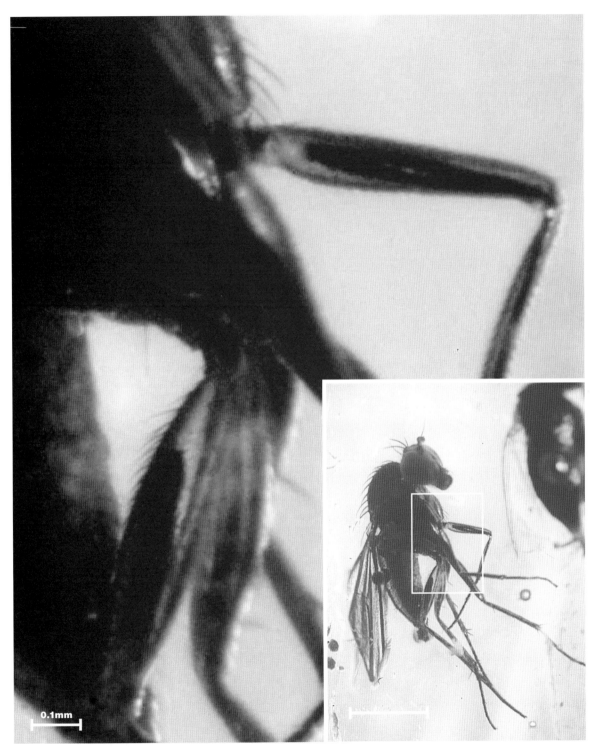

This close up of a long-legged fly in Baltic amber was made using transmitted ultra-violet light, showing the slightly shrunken muscles in the legs. An ultraviolet light was placed beneath the amber, causing the amber to fluoresce and shine through the insect. It is important not to look directly at the ultraviolet light if attempting to photograph insects in this way, as it can damage your eyesight.

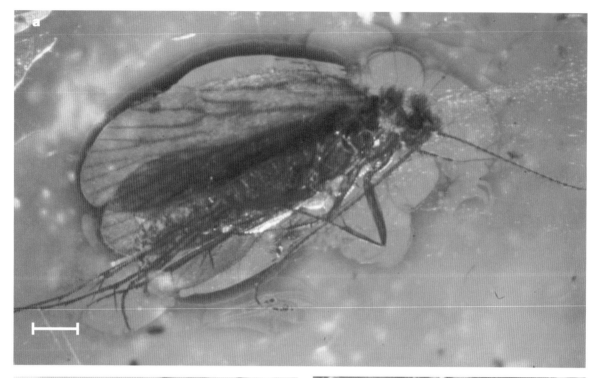

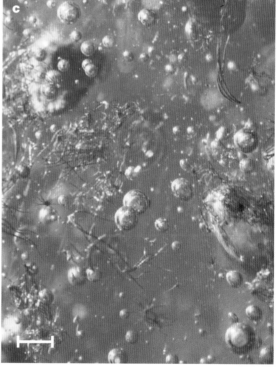

There is a 'halo' effect around this fossil caddis fly (a) possibly due to the decomposition of the fly after it was trapped. Not all inclusions are easy to identify; this strange minute shed skin of an insect larva (b) with large jaws and long antennae has yet to be identified. The most common inclusions in Baltic amber are the bubbles and stellate hairs (c). Sometimes they have bubbles of gas and fluid inside them.

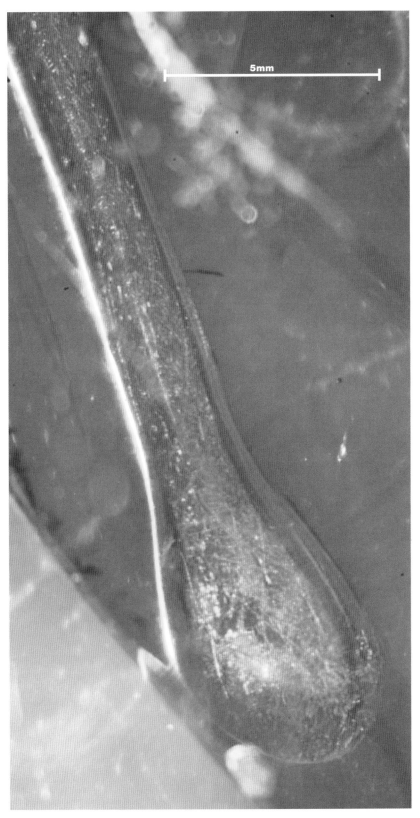

5mm

As amber drips from the resinous trees, it forms drips, droplets, or even stalactites or 'icicles'. These can be preserved as natural amber drops, but sometimes a new amber flow encloses the previous drop, trapping it like it does the insects. This picture, showing a long drip ending in a bulbous tip where the first flow ended, is an example of amber in amber inclusion. Sometimes it is possible to find insects trapped in the first flow, such as the mite trapped near the bulbous end of this stalactite.

Fakes and alternatives to amber

As a result of the popularity and value attributed to amber, there have been many attempts to produce a cheaper and comparable alternative, sometimes to deceive, and other times just to simulate. Whether or not the manufacturer originally produced the item in order to deceive, many fake items now sold on internet auctions are sold as real amber. The term 'Amber' can refer to raw amber, worked amber, processed amber, composite amber, other resins and imitation amber. Thanks to the popularity of the film *Jurassic Park*, the market for amber with inclusions has flourished. The downside to this is that there are even more falsified inclusions and fakes available to the unwary collector than at any time in the history of the amber trade.

Since the early sixteenth century and perhaps even earlier, faked inclusions have been manufactured. A frog, lizard, fish or other attractive creature may be placed into a drilled hole in the amber and then sealed. This type of fake was revealed by illustrations in publications by Daniel Hermann in 1583 and Nathanael Sendel in 1742. Although lizards and geckos have been found in genuine pieces of amber, to date there have been no fish, and this was not lost on these early pioneers of

Selection of inclusions from Nathanael Sendel's *Historia succinorum corpora aliena*, Leipzig, 1742.

fake recognition. In 1749, John Lowthorp suggested that some inclusions were products of 'art', especially those with fish inclusions. It is quite possible, however, that some genuine rare inclusions were also branded as fakes, based on the seventeenth century notion that amber was a petroleum-based resin that rose to the surface of the sea, and hence unlikely to have trapped land animals.

In 1661, a recipe for producing counterfeit amber was written by J. Wecker and R. Read.

To counterfeit Amber

You shall counterfeit Amber thus. Take chrystal beaten into very fine pouder, and whites of Eggs, excluding the Cock-Treads, and beat them, and take off the froth so long, untill they be resolved to water; mingle the foresaid pouder with them and work them together, adding a little Saffron finely poudred, if you desire to make yellow Amber; then cast all these into a hollow Reed, or some Gut, or some Glass Viol, and put them so long in scalding water, untill you find they have got a solid hard consistence, take them out, and grind them upon a Marble, and make them of what Form you please, if you will make little Beads for Ave Maries, or Hilts of Swords, make the holes before you let it dry, after that set them in the Sun; you may also mingle what colour you will, and counterfeit such precious stones as you desire handsomely. But then the matter must be strain'd before it is boyled, that the body of the Jewel may be transparent and clear. A secret of an Ingenious Artificer, a friend of mine. Mizaldus.

How to melt Amber

Amber may be handled with you're hands like Wax, and made into what forms you please, if it be cast into melted Wax skimmed. For so it will become so soft and tractable, that you may conveniently use it for Seals, or other things which you desire to make. *Mizaldus had this of an ingeneous Lapidary, and Caster.*

It is unknown how successful this counterfeit product was at fooling the people it was sold to, but as he describes it as being like 'wax', it might be fair to conclude it had a very limited market.

ARTIFICIAL AMBER

There are many products that have been used to simulate amber. The most frequently used, and sometimes incorrectly sold as amber, are copal, glass, celluloid, phenolic resins, casein, and plastics. Quite commonly amber is pressed with, or embedded in, other resins or plastics. Amber may also be treated by heating to a high temperature, to produce wonderful effects like 'sunspangles', green amber, and the clearing of opaque amber. This effect can also be produced using some plastics, making fakes very difficult to identify from this alone.

Copal is relatively young tree resin, mostly of up to only a few million years old. Because of its low melting temperature, copal frequently has insect inclusions added by unscrupulous dealers. Copal is most frequently incorrectly sold as amber, commonly from countries such as Columbia and Madagascar.

Glass has been used as a substitute for amber since the Iron Age. Some Roman and Viking hoards contain amber-coloured glass beads. Although glass is relatively easy to distinguish from amber, as it is cold to the touch, these early people may have chosen it as an alternative because unlike amber, it is robust and does not burn.

Celluloid is very difficult to distinguish from amber visually. It is made from nitrocellulose and camphor and was the first thermoplastic, created in the 1850s. Celluloid was most famously used in the film industry since the 1880s. It is easily moulded and shaped, but tends to be highly flammable, and is therefore rarely used today. When heated it gives off a distinctive acrid odour of burning plastic.

Above: Amber-coloured Bakelite handbags gifted by Martha Sabin. © Special Collections, University of Colorado at Boulder Libraries, USA. *Right:* Sun-spangles in a piece of plastic fake 'green amber'.

Phenolic resin is a synthetic thermosetting resin produced by chemical reaction Developed in the early twentieth century, one of the most common amber simulants is Bakelite, a phenolic resin mixed with wood flour. It is now rarely used, as it tends to be quite brittle. The colours of this substance can be very similar to the colours of amber and it was frequently used in the form of artificial amber beads. In the recent past, Bakelite jewellery has been fought over by inheritors who thought they had a valuable heirloom, unaware that it had, in fact, little value. Today, however, Bakelite is quite collectible and commands a good price with collectors. When heated, the phenolic resins do not smell of pine-tree resin as Baltic amber does.

Casein is a phosphoprotein found in cow's milk and commonly used as a water soluble medium for paints by artists. When combined with water and formaldehyde, it produces a yellowish compound that is slightly denser than amber. It was developed in the 1860s as the first semi-synthetic plastic, when it appeared under the trade name *Galalith*. It was later used in jewellery when a dry process of manufacture was developed. In 1915, Queen Mary was so impressed with this product that she purchased several pieces of casein jewellery at the first British Industries Fair.

Plastics such as polyester or polystyrene are still commonly used as substitutes for amber, especially in Mexico, the Dominican Republic and China. Many unsuspecting lizards, frogs and insects have found themselves encased in a plastic tomb and sold for considerable sums of money to unfortunate collectors. Many of the modern plastics are quite difficult to distinguish from amber without performing a number of tests.

HOW TO IDENTIFY TRUE AMBER

It is not always easy to distinguish amber from the many simulants, but there are a number of tests that can be carried out that will help confirm suspicions in at least some cases. When a piece of amber is already mounted in a piece of jewellery, it can be quite difficult to perform many of these tests, as some of them require that the amber be damaged a little in the process. Some tests require laboratory equipment for measuring features like specific gravity, melting point and refraction index. Others require specialised equipment such as an infrared spectrometer to measure the characteristic 'Baltic amber shoulder' that differentiates between Baltic and non-Baltic amber. These tests are of little use to the average amber collector.

Fake amber from China sold on the internet as real amber with scorpion inclusion.

The following tests can be carried out by anyone without too much damage being caused to their amber pieces.

NON-DESTRUCTIVE TESTS

Ultraviolet fluorescence

Most amber fluoresces a pale blue colour when exposed to short-wave ultraviolet light. This can show unusual and interesting flow patterns that form naturally as the amber resin accumulates on top of resin that has already solidified, or at least partly solidified. Ultraviolet light also provides a means of examining the inclusions in the amber and how they relate to these flows.

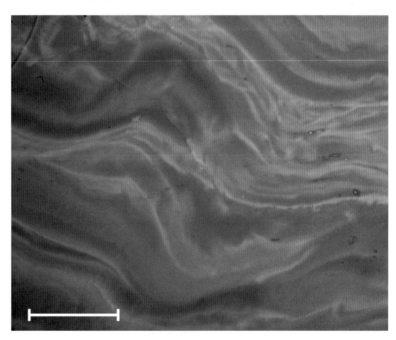

Flow patterns in amber from Borneo fluorescing under ultraviolet light.
(Scale = 1mm)

Flotation

Due to its slightly higher density, amber does not float in seawater, unless it is the very salty Dead Sea. As its density is quite close to seawater, amber is easily moved along the sea bed by currents. If amber is placed in highly saturated saltwater, then it should float. If you completely dissolve 23 gms of table salt in 200 ml of lukewarm water, this will increase the density of water enough to allow amber to float. This is a simple test, but not definitive, as some copal and modern plastics may also float. However, if it sinks, and if there is nothing attached to the sample, such as a jewellery setting, then it is highly unlikely to be amber.

Identical view of an ant in Baltic amber. a) in natural transmitted light; b) and c) in ultraviolet light. By lighting image b) from the side, the view shows the amber flow patterns, whereas c) was lit from below, showing the ant silhouetted against the fluorescing amber. (Scale = 1mm)

Identification of inclusions

This will require the help of a scientific expert such as a specialist palaeontologist (studies fossils), botanist (studies plants) or entomologist (studies insects). Most plant and insect inclusions within amber are representatives of extinct species or species that have evolved and changed slightly. If the species represented in the sample is not extinct or different in some way from its living relative, then it may not be true amber. The presence of the small stellate hairs, or trichomes, commonly found in Baltic and Mexican amber, is also a good guide to its authenticity because people who produce fakes rarely go to the trouble of adding such tiny inclusions throughout the sample. As a rule of thumb, inclusions are rarely perfect and lizards are extremely rare; however, the reverse is true of counterfeits, and the specimen is often placed in a near-perfect and unnatural position.

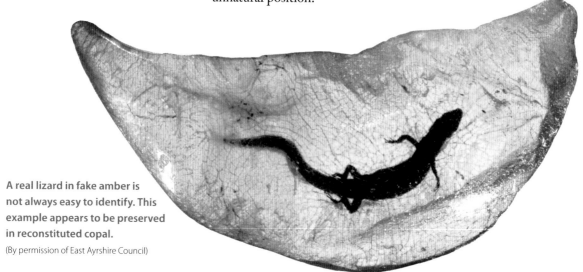

A real lizard in fake amber is not always easy to identify. This example appears to be preserved in reconstituted copal.

(By permission of East Ayrshire Council)

Polarised light

If the sample is placed between two polarising sheets and one of them is rotated through 360°, then a sequence of rainbow colours should appear. This can also be seen in some modern plastics, but what this technique does show up is any filled drill holes made to artificially introduce inclusions. Any sample that does not show the rainbow colour effect may not be real amber.

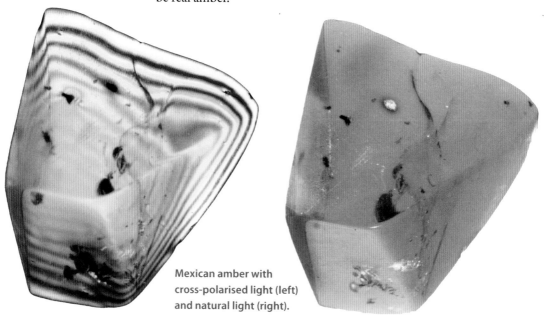

Mexican amber with cross-polarised light (left) and natural light (right).

Taste

This is an unusual test and requires a lot of experience. The sample should be thoroughly cleaned in water before carrying out the taste test. If you touch your tongue to many modern plastics, they will produce a bitter chemical taste. Amber and copal have very little taste at all.

Friction

When rubbed with a soft cloth, or in the palm of your hand, true amber will emit a slight resinous smell of pine and become statically charged. It may even be able to pick up very small pieces of paper. If this test is used on copal, it can soften and become slightly tacky to the touch.

TESTS THAT MAY CAUSE SLIGHT, OR SIGNIFICANT, DAMAGE

It is always best to try these tests on non-visible parts of the piece being tested. If it turns out to be genuine amber, you don't want to have ruined an important and valuable piece in the process!

Scratching

Using the Moh's scale of hardness, where talc is 1 on the scale and diamond 10, amber has a hardness between 1.5 and 2.5. Amber is quite soft and can be scratched by platinum, which is about 4 on Moh's scale. A fingernail cannot scratch it, however, being about the same hardness.

Melting

When the point of a needle is heated until it glows red, and then pushed into amber, the amber produces a sooty smoke. If the sample is copal and not amber, the needle will penetrate quite easily, and produce a fragrant odour. Even without heating the needle it may be possible to identify true amber, since plastic is elastic and the needle will get stuck in it. Amber is brittle and will often chip when a needle is pressed into it. Amber is also a poor conductor of heat, which is why it always feels warm to the touch.

Dissolving

One drop of acetone is dropped onto the sample and allowed to evaporate. Another drop is then placed on top of the last drop. It is very difficult to dissolve amber in acetone, and it will remain unaffected. Copal, on the other hand, will become tacky to the touch.

Cutting

If you try to shave off a small piece of the sample and it produces tiny shavings without splinters, it is likely to be a plastic. Real amber is brittle and will fracture and splinter rather than slice.

WHAT TO DO ONCE YOU HAVE A PIECE OF AMBER

Unfortunately, unlike diamonds, amber is not forever. Bronze Age amber, and even some Roman amber, has developed a thick crumbly crust. Despite the fact that amber is many millions of years old, and might be considered to be chemically stable, this is not the case. Even Baltic amber from the last few centuries has developed a thin cracked crust. Amber from the Dominican Republic may even crust over the shorter timespan of less than 20 years. In a study into the conservation of 17,000 amber items held in the National Museum of Denmark, 45% of the collections showed signs of deterioration and 25% were in immediate need of conservation.

So, if museums have problems with their amber collections, what can be done to preserve our own valuable amber jewels from the ravages of time? Until the 1980s in museums, conservation of deteriorated and deteriorating amber concentrated on interventionist approaches such as the use of chemicals. Many of the early chemicals used are now showing signs of deterioration themselves by shrinkage, cracking and discoloration which is affecting the amber that it was used to protect. A study into what causes the deterioration of Dominican amber suggests that too little humidity or too high humidity may cause damage to the amber. Oxidation of amber is also a problem, causing pits and cracks that further increase the risks of humidity damage. Generally, an attempt should be made to maintain a moderate level of humidity (50–60% relative

This amber nugget with inclusions is 8.4 cm long but only weighs 38 gms.

humidity) and a low oxygen environment. This is not something that the majority of people with amber jewellery can achieve. Perhaps the following guidelines are a bit more helpful.

1. Do not throw your amber about, as it is brittle and can fracture easily if it comes in contact with something hard or metallic. Avoid tangling necklaces, and prevent it from knocking against other jewellery.
2. Be careful when handling amber, as it is easily scratched, being the softest and lightest of gems.
3. Keep your amber away from solvents, including chlorinated water, hairspray, nail polish remover, and perfumes. The amber may react and form a thin opaque film.
4. When not wearing the amber, place it out of direct sunlight, as it will darken over time.
5. Apply a very small amount of olive oil and rub with a soft flannel, as this will help protect your amber from humidity fluctuations and oxidation. The less olive oil you use the better.

Above all, enjoy your amber, whether it is for science or aesthetics — it is a rare and exceptional gem.

Further reading and websites

FURTHER READING

General

The Quest for Life in Amber. Poinar, G. and Poinar, R. Addison-Wesley Publishing Co., 1994.

The Amber Book, translated by Jonas Leijonhufvud. Dahlstrom, A. and Brost, L. Geosciton Press Inc., Arizona, 1996.

The Northern Crusades. Christiansen, E. Penguin Books, 1997.

The Science of Jurassic Park and the Lost World or, How to Build a Dinosaur. DeSalle, R. and Lindley, D. Flamingo, London, 1998.

Amber: the Natural Time Capsule. Ross, A. Natural History Museum, London, 1998.

The Amber Forest: a reconstruction of a vanished world. Poinar, G. and Poinar, R. Princeton University Press, 1999.

The Great Book of Amber. Mierzwinska, E. Muzeum zamkowe, Malbork, 2002.

Guide to Amber. Gierlowska, G. 'Bursztynowa Hossa' Publishing House, 2003.

Amber. Gierlowska, G. 'Bursztynowa Hossa' Publishing House, 2004.

The Beauty of Amber. Gierlowska, G. 'Bursztynowa Hossa' Publishing House, 2004.

Amber in Therapeutics. Gierlowska, G. 'Bursztynowa Hossa' Publishing House, 2004.

The Amber Room: The Controversial Truth about the Greatest Hoax of the Twentieth Century. Scott-Clark, C. and Levy, A. Atlantic Books, 2004.

On Old Amber Collections and the Gdańsk Lizard. Gierlowska, G. 'Bursztynowa Hossa' Publishing House, 2005.

What Bugged the Dinosaurs? Insects, disease, and death in the Cretaceous. Poinar, G. and Poinar, R. Princeton University Press, 2008.

More specialist

Amber in Prehistoric Britain. Beck, C. and Shennan, S. Oxbow Books, Oxford, 1991.

Degradation and Inhibitive Conservation of Baltic Amber in Museum Collections. Shashoua, Y. Department of Conservation, The National Museum of Denmark, 2002.

Atlas of Plants and Animals in Amber. Weischat, W. and Wichard, W. Verlag Dr. Friedrich Pfeil, Munich, 2002.

Plant resins: Chemistry, Evolution, Ecology, Ethnobotany. Langenheim, J. H. Timber Press, Cambridge, 2003.

Amber, Golden Gem of Ages (4th Edit.). Rice, P.C. Author House UK Ltd., Milton Keynes, 2006.

Gemlore. Morgan, D. Greenwood Press, Westport, CT, USA, 2008.

USEFUL WEBSITES

http://www.gplatt.demon.co.uk

http://en.wikipedia.org/wiki/amber

http://www.amber.com.pl/eng

http://www.mhmg.gda.pl/international/index.php?lang=eng&oddzial=4

http://www.zamec.malbork.pl/en/index.html

http://www.hunterian.gla.ac.uk

Illustration credits and acknowledgements

ILLUSTRATION CREDITS

Photolibrary, Hunterian Museum and Art Gallery, University of Glasgow, University Avenue, Glasgow, G12 8QQ: pages ii, 1, 8 left, 9, 10, 11, 24 bottom, 28, 33, 34 top left and right, 59, 63, 66, 69, 76, 80, 83, 84, 86, 87, 88 top, 92, 93, 94, 95, 96, 97, 98, 99, 100, 101, 104 bottom right, 105, 106, 107, 108 bottom, 110

Kamai: pages ix, 4, 14 top, 25, 81, 91

Clare Clark: page 8 right

Malbork Castle Museum: pages 14 bottom, 15, 16, 36, 37, 38, 39, 41, 42, 44, 45, 46, 47, 48, 50, 51, 52, 53

Janina Leonavičienė: page 19

Poczta Polska: page 23

Ewa Tykc Karpińska (design)/Mennica Polska: page 24 top

Lech Okonski: page 34 bottom

Herbert Spichtinger/Corbis: page 55

Sergei Mikhailovich Prokudin-Gorskii (creator): page 56

Roland Weihrauch/dpa/Corbis: page 57

Jacek Ostrowski: page 60 left

Giedymin Jablonski: page 60 right

National Museums of Scotland, Edinburgh: pages 65, 67

Paul Tafforeau, European Synchrotron Radiation Facility, Grenoble, France: page 77

Amber Museum, Gdańsk History Museum: page 88

University of Colorado, Boulder, USA: page 104 bottom left

Dick Institute, Kilmarnock: page 108 top

Sources of other illustrations

Page 2, *Hortus Sanitatis* by Jacob Meydenbach (1491)

Page 17, *Ovid's Metamorphoses* trans. by Vergilius Solis (1563)

Page 20, *Walhall: Germanische Götter-und Heldensagen. Für Alt und Jung am deutschen Herd* by Felix and Therese Dahn (1901)

Page 31, Woodcut by Johann G. Wagner (1774)

Page 70, *Illustrated London News* 1909

Page 73, Transactions of the Geological Society of Glasgow, Volume 10 (1894)

Page 102, *Historia Succinorum corpora aliena* by Nathanael Sendel (1742)

ACKNOWLEDGEMENTS

Poland

Malbork Castle Museum: Marek Stokowski, Jolanta Ratuszna, Mariusz Mierzwinski, Elzbieta Mierzwinski, Janusz Trupinda

Guide and translator: Katarzyna Czaykowska

Muzeum Bursztynu, Gdańsk: Joanna Grążewska and colleagues

Muzeum Historyczne Miasta Gdańska: Adam Koperkiewicz

University of Gdańsk: Elżbieta Sontag

Gdańsk Artist: Giedymin Jabłoński

The Polish Culture Insitute: Aneta Prasał–Wisniewska, Natalia Gedroyć

Mennica Polska: Barbara Sissons, Mariusz Przybylski

Narodowy Bank Polski: J. Eryk Kazanowski

Scotland

Hunterian Museum and Art Gallery: Mungo Campbell, Harriet Gaston, Jill Barnfather, Mhairi Douglas, Stephen Perry, Susan Ferguson, John Faithfull, Jeff Liston, Sally-Anne Coupar, Chris Maclure, Mark Herraghty, Aileen Nisbet, Graham Nisbet, Roslyn Purss, Yvonne Threwal, Mike Richardson, Andrew Jackson, Geoff Hancock, Dave Russ, Caroline Ross, Monica Callaghan

National Museums Scotland: Andrew Ross, Helen Osmani, George Dalgleish

Integrative & Systems Biology, University of Glasgow: Craig Daly

Myra Thomson

Special thanks to Clare Clark for the discussions, editing and support

Index